To Touch the Face of God

New Series in NASA History

To Touch the Face of GOD

The Sacred, the Profane, and the American Space Program, 1957–1975

KENDRICK OLIVER

The Johns Hopkins University Press
Baltimore

© 2013 The Johns Hopkins University Press
All rights reserved. Published 2013
Printed in the United States of America on acid-free paper
9 8 7 6 5 4 3 2 1

The Johns Hopkins University Press
2715 North Charles Street
Baltimore, Maryland 21218-4363
www.press.jhu.edu

Library of Congress Cataloging-in-Publication Data
Oliver, Kendrick, 1971–
 To touch the face of God : the sacred, the profane, and the American space program, 1957–1975 / Kendrick Oliver.
    p. cm. — (New series in NASA history)
 Includes bibliographical references and index.
 ISBN 978-1-4214-0788-3 (hbk. : acid-free paper) — ISBN 1-4214-0788-4 (hbk. : acid-free paper) — ISBN 978-1-4214-0834-7 (electronic) — ISBN 1-4214-0834-1 (electronic)
 1. Astronautics—United States—History—20th century. 2. Religion and science—United States—History—20th century. 3. Astronautics—Political aspects—United States—History—20th century. 4. Christianity and politics—United States—History—20th century. 5. United States. National Aeronautics and Space Administration—History—20th century. 6. United States—Religion—20th century. 7. United States—Politics and government—1945–1989. I. Title.
 TL789.8.U5O39 2013
 629.40973'09046—dc23   2012017651

A catalog record for this book is available from the British Library.

Special discounts are available for bulk purchases of this book. For more information, please contact Special Sales at 410-516-6936 or specialsales@press.jhu.edu.

The Johns Hopkins University Press uses environmentally friendly book materials, including recycled text paper that is composed of at least 30 percent post-consumer waste, whenever possible.

For Esmé and Elias

The heavens declare the glory of God;
And the firmament shows His handiwork.
                      —Psalm 19:1

REPORTER: Do any of them have a particularly strong religious faith, or is it motivation of service to country? What are they hanging on to?
SCHIRRA: I think I should like to dwell more on the faith in what we have called the machine age. We have the faith in the space age.
                      —NASA Press Conference:
                      Mercury Astronaut Team, April 1959

The sooner we turn our eyes to Him for help, the better it will be, for ourselves, as well as for the objectives we are pursuing.
                      —Wernher von Braun, July 1965

Contents

	Acknowledgments	*xi*
	Introduction: The Blasphemy of Going Up	1
1	A Power Greater Than Any of Us: Religion and Secularity in the Formation of the American Space Program	11
2	Signals of Transcendence: The Rise and Fall of Space-Age Theology	44
3	Into the Other World: Anticipations of Spaceflight as Religious Experience	71
4	Perhaps a Meaning to Us: The Apollo Missions as Religious Experience	97
5	Evil Triumphs When Good Men Do Nothing: Religious Americans and NASA in the Autumn of the Space Age	137
	Epilogue	164
	Notes	*171*
	Bibliographic Essay	*217*
	Index	*221*

Illustrations follow page 96

Acknowledgments

There were times, some quite recent, when researching and writing this book felt a bit like being lost in space—trajectory uncertain, final destination unknown, reserves running low. Many kind and constructive external interventions were needed to keep the project on course.

The project could not have been completed without the award of a British Academy/Leverhulme Trust Senior Research Fellowship in 2007–8. The award funded a full-time replacement lecturer to take over my teaching responsibilities for an entire academic year. It allowed me to research and write two chapters of the book in draft manuscript form. Work on a third chapter was substantially advanced during an additional semester's sabbatical granted by the University of Southampton in 2008–9.

I like working in archives; I like it more when the archives are located in cities where I have friends. I am very grateful for the company and generous hospitality of George and Elena Dryden and Sam and Katie Bernet during my research trips to Washington, DC. On occasion, I interrupted family visits to the United States to track down sources that were difficult to access in Britain. My in-laws, especially William and JoAnn Morandini, not only graciously forgave such absences but also filled the resulting gaps in childminding provision.

Throughout this project, I have benefited from the assistance of many archivists and librarians: Colin Fries and Steven J. Dick at the NASA History Office in Washington, DC; Rodney Krajca at NARA Southwest in Fort Worth, Texas; Gary Spurr at the University of Texas at Arlington; Shelly H. Kelly at the University of Houston–Clear Lake; Bill Markley at the Ohio Historical Society, Columbus; David Kessler at the Bancroft Library, University of California at Berkeley; Bill Sumners at the Southern Baptist Historical Library and Archives, Nashville; and Bob Schuster at the Billy Graham Center Archives, Wheaton College, in Wheaton, Illinois. Valerie Fraedrich, of National Religious Broadcasters, was good enough to send me copies of NRB newsletters from 1975. Keith Halderman painstakingly calculated the number of signatures on a petition in favor of religion in space held in the Thomas O. Paine Papers at the Library of Congress. Robert Poole gave me a copy of NASA's official photograph of Paine with Loretta Lee Fry and her petitions. I am grateful to the High Flight Foundation for permitting inclusion of an image of Jim Irwin's proposed High Flight Lodge.

Sections of this book were read in draft form by the late Paul Boyer, Steven J. Dick, Darren Dochuk, Edward Lilienthal, Peter Middleton, Steven P. Miller, and Axel Schäfer. I am grateful for their many helpful comments. I delivered papers drawn from the manuscript at a number of research seminars (Keele, Royal Holloway, Sheffield, and Southampton), as well as at conferences in London (HOTCUS), New York (OAH), Nottingham (BAAS), and Bristol (Heavenly Discourses). Frequently a question asked on these occasions prompted me to rethink some element of my approach to the subject. Less formal exchanges with colleagues at Southampton, particularly Neil Gregor, Mike Hammond, John Oldfield, and Mark Stoyle, and with my coconvenors at the Institute of Historical Research's American History Seminar in London were always sustaining and productive. Anonymous readers for the Johns Hopkins University Press rightly insisted on rigor, clarity, and coherence throughout. The history editor at the Johns Hopkins University Press, Robert J. Brugger, delivered brisk and priceless advice and encouragement whenever I needed it, which was more often than he may have wished. Joanne Allen's acute editing corrected many shaming infelicities and also greatly sharpened my prose. Kara Reiter was reassuringly helpful and efficient as the book made its way through the production process. All remaining faults in the book, of course, are my responsibility alone.

Finally, to family. My parents, Terence and Christine Oliver, were, as ever, an essential source of love and good cheer as well as excellent food and wine. My wife, Patricia Morandini, kept her faith in the project even as mine began to waver; she put up with my absences abroad (or even just upstairs) when there were two small children to look after and she had her own work to do; and when I returned from the archive or ended my studies for the day, she made me truly happy that I had done so. I could not have written this book without her. Our daughter, Esmé, was born as I was embarking on my study of religion and spaceflight; our son, Elias, appeared in time to delay its completion. I would not have wanted it any other way. Children hold us in the moment but also fold us into the infinite. The book matters; it does not matter. I hope that one day Esmé and Elias will try to read the whole thing, but all they really need to do is look at the dedication and know that their daddy wrote it with love.

To Touch the Face of God

INTRODUCTION

The Blasphemy of Going Up

"You must see it as a prayer because when we launch that thing it's like praying."[1] It was late May 1964, and the NASA astronaut Theodore Freeman was speaking to his friend Oriana Fallaci as they stood by the swimming pool at the Holiday Inn, Cocoa Beach.[2] The next morning, a short distance away at Cape Canaveral, a Saturn I rocket was due to rise from its pad and carry into orbit a boilerplate mock-up of the Apollo spacecraft system that was intended, in time, to convey man to the moon. If the launch was successful, then NASA remained on course to achieve a lunar landing by the end of the decade, a deadline that had recently acquired a new solemnity with the death of the president who had proclaimed it to the world.[3]

Fallaci, an Italian journalist, had traveled to the United States with the intention of discovering the true nature and meaning of events like the Saturn launch. "Where will the space race bring us?" Fallaci asked. "Where are we running and why?"[4] There was some seed of significance, then, in the slippage in Freeman's phrase: would the launch of the rocket be a prayer, or like a prayer? Was this just the most modern way for man to speak in praise of God, or might he whisper to himself, under the roar of the rising Saturn, that he was now so accomplished that his need for God had passed? Fallaci, as she began her work, did not know what she wanted the answer to be. She was profoundly ambivalent about the Old World, from which she had come, unable to entirely shake off the mental freight of its sacred history, her childhood fear "of angels, saints, the Virgin Mary, Child Christs and Christs Crucified, of Paradise, Hell and Purgatory, of what they call Good, and what they call Evil."[5] Perhaps the space age represented a means of final liberation from this tenacious medieval imaginary. It also offered an alternative promise of immortality, for when the sun eventually died, men might use their rockets and capsules to escape to some other solar system, carrying with them the gift of life and the best of their culture. Conceivably, however, that promise would turn out to be a curse. Arriving in space-age America, Fallaci encountered an artificial world—plastic grass on the lawns, glass flowers in a vase—and fanatical technicians who declared the superfluity of Shakespeare. Shut inside an Apollo spacecraft, she "felt caught in a trap, resigned to death."[6] Of the original Mercury astronauts, the one she liked most was Deke Slayton, who had not been allowed to fly. Alan Shepard made her angry, calling spaceflight a "commercial enterprise," and there was, she decided, something too deliberate about the charm of John Glenn.[7] Also, for the daughter of a resistance hero, there had been too many German hands involved in the making of the Saturn rocket for her to invest her faith too readily in the future it would bring.

By early 1965, when Fallaci concluded her inquiries, the book that she was writing—which she described as "an obsessive swing between yesterday and tomorrow"—had resolved itself finally into a conversion narrative, addressed to her father back in Italy.[8] She had met a German rocket scientist and found him kind and friendly. She had visited an Apache reservation—a nakedly "commercial enterprise," peddling souvenirs of its antique heritage even as its customs and values were allowed to decay. She had witnessed the astronauts recite Shakespeare from memory. Then there was the Saturn, "an enormous candle waiting to be lighted to the glory of ourselves, not of God." It swayed on lift-off, "as if it couldn't face the dare, the blasphemy of going up," but slowly it rose with a roar that reminded Fallaci of the bells at Easter, "the Easter when we are happy and free and good."[9] She moved to New York, commuting often to Houston or to the Cape, making her home in the space age. And finally there came the news, as she visited the factory where the lunar module was being built, that Theodore Freeman had died after his airplane collided with a wild goose. At his funeral, all the other astronauts coolly told Fallaci the same thing: that it could have happened to any of them. They accepted the chance of death as the price to be paid for living in the future, for the privilege of claiming the cosmos as their own. She, in turn, accepted the lesson that

> Heaven doesn't exist, Hell doesn't exist, goodness doesn't exist, but life exists, and continues to exist even if a tree dies, if a man dies, if a Sun dies. Didn't you believe it too, Theodore? I do, now. I do. And you must too, Father. Believe it, please, Father. Believe it together with me. Don't leave me alone to believe it alone with them. They've convinced me, they've bent me, they've converted me, they've made me join them, and they frighten me so, Father. Because reason is on their side. And reason is always so frightening.[10]

It might have taken considerable courage to say it to her face, but within a matter of years Fallaci's fervent conversion to the space age had come to seem somewhat quaint. By the end of the 1960s, even as the Apollo astronauts took their first steps on the moon, man's adventures in the cosmos were no longer held to be so definitive of their time. For many Americans, they represented a distraction from more important concerns: the war in Vietnam, the plight of the poor, inflation, the environment, and the state of the nation's cities. A similar assessment is evident in most historical surveys of the period. Historians have rarely preferred *space age* over other synonyms for the "long Sixties," tending to reach instead for the language of conflict and dissolution. After Kennedy's "Camelot" was ended by an assassin's rifle, there followed "years of discord," of the "unravelling of America," even of "civil war"; the culminating drama was Watergate, which began with an "enemies list" and concluded with a president's disgrace.[11] In such accounts—of struggles across (and as often within) the boundaries of race, generation, gender, and ideology—the missions to the moon sometimes are not mentioned at all.[12] And when they are, they are often cast as aberrations, rare, fleeting moments of national unity and optimism, after which everyone went back to arguing about Vietnam.[13]

That the space age appears to offer an unsatisfactory narrative framework for our current accounts of the long Sixties may be owing at least in part to a problem of chronology. The entry of a man-made satellite into the heavens, or more precisely, the response it provoked in terms of new state-directed ventures in technology, education, and poverty reduction in the United States and other Western countries, still plausibly marks 1957 as a watershed of sorts.[14] Dating the end of the space age presents more of a challenge. Indeed, a good case can be made that it never actually ended. Throughout the 1970s and 1980s, and up to the present day, the United States and the Soviet Union (now Russia) have continued to dispatch manned and unmanned missions into space, and they have been joined there by Europe, China, Japan, India, Iran, Israel, and the Ukraine, as well as by a handful of nonstate actors. Over the same period, the interpolations made in our lives by space technologies, most obviously satellites, have steadily increased in diversity and number. "We must ponder the ambiguities of the Space Age," noted the historian Walter McDougall in 1985, "because we have to live in it, willy-nilly."[15]

Yet for other commentators, notably the British writer Marina Benjamin, the space age is most assuredly over, having met its demise about the same time as the rest of the long Sixties. The launchings would continue, and space enthusiasts would go on dreaming, but for everybody else the shutters had long since come down, the contents hauled off to be interred in a museum: "Today, when we think of the Space Age, we think of the Cold War and the birth-control pill, of Elvis and flower power, of a time-warped era sealed off from the present, for ever."[16] Benjamin's argument is a reasonable one. The space age was not defined by the fact that man and his inventions were now journeying beyond the earth's atmosphere, but by the cultural optimism that attended their travels, by high-modernist expectations that technology and technocracy, once their relations had been perfected in the skies, would become the midwives of social progress back at home, and by the romance and promise of exploration and contact across an endless interstellar frontier. The space age was indistinguishable from the hopes invested in it—those of Oriana Fallaci and millions of others. But a withering of those hopes was not difficult to detect even before the launch of the last manned lunar mission, *Apollo 17*, in December 1972. If, in the roar of that rocket, bells could be heard, then they were tolling for the space age.[17]

That the US space program, in its Mercury, Gemini, and Apollo iterations, no longer seems to define its era in the way it once did is a function of content as well as chronology. Not only is the space age sealed off from the present but it was also, in a sense, detached from its own time. After the 1960s, partly in protraction of the struggles of that decade, the historical discipline directed its attention to the study of minority or subaltern social groups, seeking to accelerate the redress of generations of political neglect by ending at least the condescension with which these groups had been viewed within the academy. Many such studies took as their subject the decade of the 1960s itself, with its pageant of rising social movements, but they found little of relevance to their concerns in the story of America's enterprise in space. There may be a social history of NASA and its networks of installations and contractors to be written, but it would be one from which women and ethnic minorities were largely excluded until the late seventies.[18] Moreover, NASA's managers showed little sustained interest even in environmentalism, the one social movement that the agency might plausibly be said to have inspired, by providing a series of poignant images of the earth from outer space.[19] In the literature on NASA in the Apollo era, there are recurring references to the pressurized suits worn by the astronauts; the sterile White Room, through which they processed at the Cape Canaveral launch pad; the systems in their spacecraft that protected them from the vacuum outside and from its extremes of heat and cold; and, on return from the moon, the Mobile Quarantine Facility and the Lunar Receiving Laboratory, in which they were confined. It is difficult not to see these features of space exploration as symbolic of a more fundamental condition of insulation and isolation, as if the space program existed in its own separate decade, apart from the rest of the 1960s. Indeed, there were probably few institutions in the United States as isolated as NASA was from so many of what historians have considered to be the principal social and political currents of the time. "You didn't care whether they were marching in Mississippi," remembered the *Apollo 12* astronaut Alan Bean. "We never talked about anything but space."[20]

This is a theme that can be carried too far. Even though the core professional workforce at NASA centers and their contractors might have appeared almost comically homogeneous and monastic in their self-absorption, the space program undoubtedly made its mark on the world beyond and was marked by it in turn. Research into the space age need not be confined to the ghetto of official histories, narratives of hardware development, and endless popular retellings of how the roaring boys of NASA's astronaut corps burned a trail to the moon. That was the lesson of the original classic of space history, Walter McDougall's . . . *the Heavens and the Earth*, with its account of how the space race transformed conceptions of the proper relation between state and society in the United States and elsewhere. In the wake of two further formidable studies, Howard E. McCurdy's *Space and the American Imagination* and Robert Poole's *Earthrise*, alongside a number of agenda-setting essays, perhaps we can even identify the coalescence of a "new aerospace history."[21] McCurdy examined the efforts of American space enthusiasts from the 1940s onward to align their manifestoes with longstanding themes in the country's culture (such as the frontier) and to create a vision of cosmic exploration that was sufficiently popular and compelling to influence public policy and convert the moon into a national goal. Poole, in contrast, traced the ironic history of the Apollo missions themselves. Of the new perspectives that they yielded, the most compelling actually proved to be the distant view of a fragile earth, which drew public attention away from the heavens and back in the direction of home, extinguishing the outward thrust that had qualified the sixties as the age of space. By revising the productivist bias of aerospace scholarship and engaging with modes of consumption and reception, both authors sought to reconnect the space program to the ideas and concerns that have animated the wider profession of academic history over the past thirty years and to the rest of what happened in America between Sputnik and the last launch of a Saturn rocket in 1975.

To consider these two works together—one lingering on the American romance with space and the other disclosing how the affair came to an end—is to grasp the challenge for aerospace history once it departs from its heroic, exceptionalist mode. Certainly there is a need to account for the Mercury-Apollo space program as a signature phenomenon of American society and culture in its own time, as an agent of social change, as a stimulus to cultural production, and as the focus of hopes and fears about man's future in the cosmos. But any attempt to claim for that program a truly transformative role eventually must contend with the knowledge that so many of its principal managers and advocates ended up disappointed men, sorely feeling their marginalization within a culture that had moved on.[22] Their prodigies of invention had reached for the starry skies without sinking more than the shallowest roots down into their native soil. What measure of transcendence had they really achieved if by the early seventies they were mired in battles to justify the costs of their ambitions against the commands of earthly priorities and if public acclaim for their spectaculars had dwindled away into indifference? The space age was constituted from an unstable compound of popular attraction, ambivalence, and unconcern. The overall balance among these attitudes lay beyond NASA's control, for it was not just determined by the agency's record of achievement in space. How Americans regarded the Apollo enterprise was influenced by a range of variables external to the program, most obviously the state of the national economy and the temper of Cold War relations but also, at a more granular level, the local circumstances in which they lived and worked and their own individual habits of thought and belief. Investments in hardware to be sent to the moon could look very different to the residents of America's impoverished city centers than they did to the suburban aerospace engineer surfing a long spring tide of NASA-funded R&D. Man's excursions across the lunar surface may have fascinated many amateur astronomers, but what about participants in the counterculture, with their emphasis on the exploration of *inner* space?

Recently, of all the major social variables, religion has probably attracted the most conspicuous attention from students of the postwar United States.[23] Their interest in religion is conspicuous because it is new. Only a few years ago, in a seminal intervention, a leading scholar observed that the historiography of modern America tended to conform to the predicates of the secularization thesis.[24] Religious convictions were assumed, if not actually to be dying out, then at least to be losing their salience across the public life of the nation. When religious individuals and movements made an appearance in historical survey texts, either they were cast as somewhat out of time, like the fabled Japanese soldiers who maintained their vigil in the jungle for years after the war was lost, or else their essential motivations were described in secular terms. Outside the subfield of religious history there was little serious and sustained scrutiny of religious faith and its work in the modern world. And this was true of most accounts of the space program in the Mercury-Apollo era, which seemed implicitly to accept Fallaci's proposition that men who had fashioned the means to deliver themselves unto the stars had no real need to trust in the care of God. The standard histories of the program at best gesture briefly in the direction of the faith of John Glenn, the reading of Genesis by the crew of *Apollo 8*, Buzz Aldrin's celebration of communion after the first lunar landing, and Jim Irwin's epiphany of God's presence as he walked on the moon and then return to their secular themes: the Cold War competition with the Soviets; the development of spaceflight technology, concepts, and technique; the astronauts' encounters with risk; the scientific yields; and finally, more painfully, the termination of the endeavor in an era of economic limits.[25] Only Robert Poole, with his interest in how the program revised mankind's apprehension of its cosmos, has offered much more than vignettes of astronaut piety and surveyed the intersections of spaceflight and spirituality from a wider angle of view.[26]

Conversely, but with the same consequence, few scholars of postwar American religion have really known what or even whether to think about the nation's enterprise in space. In most accounts of religion in the sixties, the space program either is not mentioned at all or features only briefly, popping up at odd moments in the narrative to show the colorful ecumenicity of religion's encounter with the modern world during the course of the decade.[27] The reader may be told of the Methodist bishop who urged his church's seminaries in 1960 to start preparing their students for conversations with intelligent beings from other planets, of the *Apollo 8* reading of Genesis (again), and of the dream of cosmic rebirth that concludes Stanley Kubrick's *2001: A Space Odyssey*.[28] These are discursions, though, not much more than marginalia; the real action, it is evident, lies elsewhere, with Vatican II and its impact on American Catholicism, with the religious motivations, language, and rituals of the civil rights and antiwar movements, with the rainbow spiritualities of the counterculture and the nascent New Age, and with the surprising muscularity of conservative evangelicalism as the long Sixties neared its close.

Such neglect seems strange. The underlying supposition of scholars of postwar religion was that their subject, defined as a motivating faith in a largely unseen general order of existence, retained its relevance amid all the cool, empirical instrumentalism that marked the condition of modernity. Where better, then, to test the issue than the space program? For David F. Noble, the development of new technologies has almost always expressed a hope of redemption, of recovering the divinity that man forfeited in the Fall.[29] Spaceflight, specifically, conjured with the symbolic tradition that redemption and divinity were manifested in motions of ascent toward heaven. The point was proved, he thought, by the presence of committed Christians across the ranks of NASA employees. Whereas the secularization thesis anticipated a technocratic venture untouched by transcendent concerns, Noble took the view that transcendent concerns, in actual fact, had directed the program's goals and continued to inspire its personnel. He was not inclined to celebrate the theme. More than any other technology, he judged, the machinery of spaceflight involved an explicit, and immoral, evasion of earthly priorities for the sake of chasing phantoms of the sacred beyond the sky.

Did Americans really conceive of spaceflight as a means for mankind to return to God? Or was Oriana Fallaci right to identify the launch of a rocket as a blasphemous act, an expression of man's growing conviction that no metaphysics could constrain his mastery of the cosmos? Or was it possible that religion neither inspired the nation's space program nor perceived it as a threat, indeed that these two realms of human activity, for all their shared interest in the heavens, did not speak to one another in any significant way?

For both histories of postwar American spirituality and accounts of the space age the answers to these questions matter. To examine the relationship between religion and the space program is to disclose something important about the continuing capacity of conceptions of the sacred not just to hold their own in modern technocratic contexts but actually to shape behavior and participate in the process of ascribing meaning to technological innovation and the new experiences it makes accessible. Indeed, to the extent that spaceflight was a charismatic enterprise, with a cultural appeal that exceeded its correspondence with calculations of national interest, it owed much to religious archetypes and sensibilities. There was, for example, an analogue of reverence in the celebrity enjoyed by the early American space pioneers, especially in the wake of a successful "mission": like the Catholic saints or the Puritan elect, these were men who seemed to be fully possessed of grace. Did such consonances make the critical difference in converting a space program into a space age? If they did, then explanations for why the space age came to an end must take account of religion too. It is a commonplace of the scholarship that by the mid-1970s many Americans had become "disenchanted" or "disillusioned" with their nation's activities in space. The terms imply that a spell had been broken, that a promise had gone unfulfilled, that a community of meaning had dissolved. Did the space age end not only because spaceflight no longer seemed so instrumental to the interests of the state but also because it had failed to deliver the hoped-for spiritual returns?

This book tells the story of relations between religion and the US space program by focusing on four major themes. Firstly, it considers the question of motivation, exploring the role played by religious faith and values in the work of the pioneers of aviation and rocketry, in the development of national space-exploration goals during the 1950s and 1960s, in NASA's institutional culture, and in the careers of American astronauts in the Mercury-Apollo era. The balance of the secular and the prophetic in the identity of the space program has salience for the book's second theme: the implications of spaceflight for religious thought and belief. Mankind had always looked to the skies for intimations of the divine. As those skies opened up to human exploration, theologians and lay commentators wondered whether man's abiding conception of heaven, his understanding of his own dependence on God, and his belief in the special place of the earth in the scheme of creation would be challenged or reaffirmed.

Thirdly, and more elemental than the abstractions of religious doctrine, there was the prospect that travel into space would admit new opportunities for spiritual experience, not just for the astronauts but also, through the realtime technologies of mass communication, for the watching public at home. Americans could follow as their heroes ventured into the traditional domain of the gods, enduring sensations of unprecedented machine force, subsisting in conditions of prolonged isolation, encountering unknown environments. If the astronauts were transformed as a result of their experiences, there was the chance that secondary effects would be registered among the audience on Earth. The space age was built in part on anticipations such as these. Conversely, if the astronauts were not transformed, public interest in the project of human spaceflight could conceivably subside, for why bother with a cosmos if it left no mark on a wayfarer's soul?

By the time of the first moon landing, the question of whether the space program should function as an auxiliary of religious faith or tend to default to secularity had become highly controversial. The final theme of this book is the way in which the survival of a spiritual dimension to the program became a subject of organized concern at the religious grass roots. In a correspondence campaign on a scale without precedence in the postwar era, millions of ordinary Americans of faith sought to persuade NASA to maintain itself in a relation to God. It was a campaign with implications for the future of the space age. At a time when the program was short of natural allies in its effort to push on from the moon, here was a mass constituency disclosing a basis on which its support could be secured: in Theodore Freeman's terms, if spaceflight truly became a prayer.

There was something about spaceflight that seemed consonant with sacred purposes; there was also something that seemed to anticipate a future, desacralized world. The space program was culturally significant because it involved participants and audience alike in a discourse of ultimacy that simultaneously revealed the influence of religion and raised the prospect of its negation. This dualism not only charged American spaceflight with much of its meaning during the space age; it also shaped the trajectory of the space age itself. The commerce between religion and the space program presented each with an existential test. Could faith in the supernatural sustain itself in the face of mankind's efforts to naturalize the heavens? Could the program flourish on a basis of material achievements alone? What follows is the tale of those tests and their outcomes.

CHAPTER ONE

A Power Greater Than Any of Us

Religion and Secularity in the Formation of
the American Space Program

In the summer of 1969, a few weeks before he was to depart for the moon, Edwin E. "Buzz" Aldrin, the pilot of the *Apollo 11* lunar module, made a request of his minister, the Reverend Dean Woodruff, of Webster Presbyterian Church in Webster, Texas.[1] Dissatisfied with the narrowness of the historical perspectives that were informing public discussion of the lunar landing mission and its significance, Aldrin asked Woodruff "to come up with some symbol which meant a little bit more than what most people might be thinking of. What we were looking for—what I was interested in—was something that transcended modern times." Woodruff prepared a paper entitled "The Myth of Apollo 11: The Effects of the Lunar Landing on the Mythic Dimension of Man." The journey to the surface of the moon, he asserted, would have an impact on "man's self-understanding" equal to that of the Copernican revolution and the publication of Charles Darwin's *On the Origin of Species*. Woodruff drew in particular on the account of human mythology offered by Mircea Eliade, professor of the history of religions at the University of Chicago. Eliade had concluded that the motif of "magic flight" was a universal feature of ancient religion and myth: it was evidence of a "longing to break the ties" of human bondage to the earth that was actually "constitutive of man."[2] According to Woodruff, the principal significance of *Apollo 11* lay in its material fulfillment of this perennial dream of escape and, more generally, in its potential for rescuing modern man from his condition of provincialism and relativism and restoring to him the consciousness of cosmic myth and symbol that had grounded the societies of his ancient ancestors. Woodruff wrote: "Perhaps when those pioneers step on another planet and view the earth from a physically transcendent stance, we can sense its symbolism and feel a new breath of freedom from our current cultural claustrophobia and be awakened once again to the mythic dimension of man."[3]

Here was the American astronaut, convinced that the fall of his foot on the shores of another world would have a meaning that extended beyond the horizon of modern history. Uncertain what that meaning was, however, he turned to his pastor for aid. And here was the pastor, reaching not, as might have been expected, to the canon of Christian apologetics in his effort to define a cosmic purpose for space exploration but to the field of comparative religion and to the descriptions it had provided of pre-Christian symbols and myths. And here too, evoked as authority, was perhaps the most eminent religious scholar of his day. Symbolic thinking, for Mircea Eliade, was an enduring and universal attribute of human existence, but in modern society it tended to persist in degraded, laicized forms.[4] It was an abiding theme of Eliade's writings that if modern man were to experience a spiritual renewal, he would have to work back from these vestiges to the complete mythologies that had provided his forebears with a coherent paradigm of being.[5] Had Eliade turned his attention to human spaceflight (interestingly, there is no evidence that he did), he probably would have found it unremarkable that, as an exemplar of modern man, Buzz Aldrin both sought to make symbolic sense of his adventure and yet did not know how to begin.[6] As the authority for the move, Eliade would also have discerned a compelling logic in Woodruff's turn toward pre-Christian religious tradition for the source of the impulse toward flight, for Christian symbols of ascent were largely derivations from archaic myth.[7] Moreover, as we shall see, Christian doctrine itself offered no clean, simple stimulus to mankind's designs on the skies.

It is the question of a religious motivation behind the American movement into space that I address in this chapter. The question is substantive for two reasons, the first historical, the second sociological. Replicating the dualism of an enterprise that geared technocratic rationality to a vault into the heavens, the historiography of the US space program is constituted in the main either by works that default to a secular tale of secular men striving to attain a shared secular goal or else, in the sharpest of contrasts, by accounts that represent it as led and staffed by the pious and inspired in its purposes by Christian millenarianism. David F. Noble states: "Religious preoccupations pervade the space program at every level, and constitute a major motivation behind extraterrestrial travel and exploration."[8] In neither of these characterizations—a space program that was essentially propelled by secular interests and concerns or one that was vitally expressive of religious intents—has the evidence for the alternative view received conscientious attention. And in both, the power to explain the emergence of spaceflight as a human aspiration and then as national policy is constrained by reliance on chronically underdeveloped taxonomies of motivation and cause.[9] Before their influence on the history of the space program can be reasonably asserted or discounted, motivations that begin at a point of religious conviction—or, more broadly, in a conception of transcendent value—must be examined in situ alongside those that do not: instrumentalist professional and institutional logics, an intrinsic commitment to exploration, engineering or spaceflight itself, even the demands of the individual ego. In addition, there is a need to distinguish between motivations that explicitly identify spaceflight as the necessary arena of activity and motivations that, though applied in this instance toward ventures in space, could as readily find fulfillment in another line of work. Was the providential purpose to be satisfied by the space program alone, or would any advance in American power or human technology and knowledge suffice?

This chapter takes the claim of a formative religious influence on the American space program seriously enough to test it against the historical grain. It is intended, though, to offer something more than just an affirmation or refutation of a single proposition. In the late 1950s and early 1960s the space program, as the exemplary product of American modernity, seemed to define its era. To the sociology of religion in the postwar United States, it presents itself as a potentially instructive case. For much of the twentieth century, reflecting perspectives inherited from the principal founders of their discipline, sociologists took the view that modernity was corroding the traditional patterns of religious practice and faith; its acids were the catalyst for the secularization of the world.[10] Later, in the wake of the 1970s evangelical revival and survey returns that indicated that Americans had refused to dispense with their belief in God, the secularization thesis was widely regarded, even by those who once had been its advocates, as a prophecy that had failed.[11] Yet the persistence of religious faith did not confute every account of how secularization would make itself manifest in the modern age. In other versions of the thesis, secularization might be evidenced not so much in the dissolution of religious faith itself but rather in the diminishing salience of that faith to activities and thought in arenas—politics, work, leisure—located beyond the physical perimeters of church and temple. Remarkably, neither this concept of a differentiation of spheres nor its obverse—the assumption that religious belief, where it exists, cannot help but direct how a life will be lived in all of its aspects and locales—has been subjected to thorough and comprehensive sociological investigation. In the postwar era, the few studies that were conducted returned ambiguous results: the influence of religious faith was certainly discernible in the broader attitudes and behaviors of those who professed it, but it rarely seemed determinative.[12] To examine the role of religion in the Mercury-Apollo space program, then, is to inquire into the substance of secularization, taking a sample from the very marrow of modernity. It is not enough to assert, as some commentators have, that many NASA personnel—officials, engineers, and astronauts—were religious believers.[13] It matters also whether their beliefs were actually relevant in their work.

Evidence of differentiation in the context of a modern, technocratic enterprise such as the space program would not necessarily establish that modernity itself was the source of secularization. Secularization theorists were heavily indebted both to the insight of Emile Durkheim that the origins of religion lay in the need of the social group to create a collective representation of itself and to the systemic models of human culture developed in the field of social anthropology.[14] What resulted was the presumption that religion, in its elementary or ideal form, was seamlessly integrated into every aspect of human activity and thought. The presumption was further enforced for those theorists who viewed the transcendental claims of religion as critical to its cultural authority and reach: only a transcendental power, after all, can make a universal law. Thus the "classical task of religion" was to construct "a common world within which all of social life receives ultimate meaning binding on everybody."[15] Hence also the conclusion, when it was intuited that religion in the modern age might fail the test of universal salience, that its deficiency of reach was something rather new and constituted a sign that a secular world was being born.

Yet the archives of anthropology also provided evidence that man in his "primitive" state was indeed able to differentiate between a realm of the profane, in which everyday practical activities were governed by an empirically acquired knowledge of cause and effect, and what Bronislaw Malinowski termed a "sacred region of cult and belief."[16] In archaic societies, Malinowski argued, the religious and the collective were not necessarily coextensive.[17] Moreover, as Durkheim himself had observed, the division of the world and everything it contained into either the domain of the sacred or the domain of the profane was the one characteristic that all religions had in common.[18] It was in the elemental nature of a religious system to assert the existence and define the parameters of a space in which some things at least were not to be rendered unto God. However, most scholars were most interested in the mind and practice of man when he entered the domain of the sacred. What happened when he returned from his rituals—what he carried back with him to the world of the profane—retained, rather ironically, the greater element of mystery.[19] Thus, when the space-age technocrat left church on Sunday morning and redirected his thoughts to the practical challenges of his work, his situation was hardly novel; nor, however, was it very well understood.

Identifying the presence of a religious inspiration in ostensibly secular activity is no simple historical procedure. Even where an individual makes the claim of a personal motivation from faith, some analytical discretion is still advised, for we do not always know, or want to know, what moves us, and in what ratios and combinations. Amid all the public piety of the American 1950s, professing a religious vocation may have been an act of social pragmatism rather than a marker of genuine commitment and belief. Moreover, what might be taken as evidence of religious influence within the secular realm—for example, the use of liturgical language or the mimesis of Christian rituals—could be interpreted another way. Habits of speech and gesture that have devolved from sacramental practice may subsist within the secular culture precisely because their religious meanings have become degraded. That these habits of speech and gesture make an appearance from time to time in the history of the US space program is not conclusive proof of its ultimate purposes: they may be secularizations of the sacred, not sacralizations of the secular. Yet it is possible too that they do constitute a clue to an original millennial inspiration, an inheritance that technology, for all its modern secular applications, cannot escape entirely.[20] The agnostic astronaut, like Weber's capitalist in his iron cage, may still be enacting an enterprise invented in the minds of saints.[21] And what counts for more, the personal, secular motivations that placed that astronaut atop his rocket or the religious hopes that, centuries past, set the direction in which he travels?

If this task of analysis is challenging, it is hardly uniquely so. The broader causative effects of religion are often diffuse and highly mediated, but so are those of other historical variables, such as race, class, and gender, which have lain at the center of our discipline's research concerns for much of the last half-century.²² Moreover, if historians of modern American religion are finally to refute the contention of secularization theory that the horizons of their subject have receded to the precincts of churches and the private reveries of individual believers, their energies must be directed precisely to the labor of determining where else, and in what manner, religious faith has made its mark.²³ The question of what the space program owed to a religious aspiration also mattered to Americans during the space age itself, for it was not always clear to them why their nation was embarked on such a venture and why they themselves were cheering. What reason could not explain, religion might. The question was presented in Saul Bellow's novel *Mr. Sammler's Planet*, set in New York during one of the lunar missions: were the Apollo spacecraft just like cars—"Machines for going away"—or did their journeys speak of some necessity of the soul?²⁴ Perhaps, as Norman Mailer concluded, voyages in space were the only means left for modern technological man to transcend his own technique and feel once again the press of metaphysics on his will.²⁵

Technology

The technological revolution in the West, famously asserted the historian Lynn White, began with the Benedictine rule. In the classical tradition, a moral distinction had been made between the work of the body and the work of the mind. But for the monks of St. Benedict, to labor was to pray, and thus their intellects could turn to the practical problems of power and production in fulfillment of their calling rather than as a diversion from it.²⁶ In addition, technological innovation promised the completion of man's dominion over nature and so a hastening of the day when he would recover from the Fall and be restored to grace, as a creator in the likeness of God.²⁷ In that sense, Weber's Puritans, with their ethic of productivity, were simply improvising on a well-established theme: in industry lay good conscience and a prospect of the blessings of heaven. Into the nineteenth century, and particularly in the United States, similar hopes of imminent social perfection were invested in the industrial revolution and the religious revival, and for some commentators, at least, those hopes were combined: through technology, Americans could convert their country into the site of a second creation.²⁸ An affinity with technological culture has continued to be evident throughout the ecologies of modern faith: many committed evangelicals work in high-technology industries; the churches to which they belong purposely coopt the latest innovations in communications systems in order to assist dissemination of the "good word"; and male spirituality more generally has been marked by a practical can-do ethos held in common with industrial labor and management.[29]

But though they may still embrace modern technology, Americans of faith no longer swoon with the millennial optimism of old. By the late Victorian age the hope that a technological utopia might be constructed in the United States was fading rather quickly: what providential purpose was to be identified in protracted spasms of industrial strife and the degradation of the urban poor?[30] "The promised land flies before us like the mirage," wrote the reformer Henry George. "The fruits of the tree of knowledge turn as we grasp them to apples of Sodom, that crumble at the touch."[31] In the twentieth century, religious writers could readily conceive of technology as a chariot to perdition, to a man-made hell in which utilitarian values were sovereign, rather than as a means of speeding redemption.[32] Moreover, the industrial division of labor was identified by secularization theorists as one of the principal sources of the social pluralism and differentiation that had operated to divest religious traditions of their aura of necessity. Men and women in industrial society were pressed into encounters with those who had been raised within faith worlds that were sometimes very different from their own.[33] In industrial environments, which were increasingly specialized, it was the injunctions of management or vocational codes and routines, not the transcendental claims of the church, that seemed most salient to most workers most of the time. By the late sixties and early seventies, jeremiads that detected in technology a remorseless pathogen against moral and spiritual values could draw on sociological studies that indicated that out of all the modern professions, it was engineering that had most clearly failed to establish a philosophical basis for its own practice. Engineers, these studies asserted, tended to be narrowly educated, careerist, and only shallowly integrated into the broader communities in which they lived. They had quarantined themselves from the sort of social and cultural experiences that prompt and sustain an interest in fundamental questions of purpose and meaning.[34] Even for those who questioned such typologies, who pointed to the spiritual gratifications that engineering could often provide, the model was not so much the monks of St. Benedict, conceiving their labor to be a prayer to God, but the Zen Buddhist with his quiet, careful reveries: the engineer, utterly absorbed in his machine and in the rational, ordered process by which he will make it work.[35] Technology may have owed its early development to a belief that increasing the power at men's hands fulfilled a providential purpose. By the dawn of the space age that belief, where still expressed, was vigorously contested, and many of those involved in technological innovation went about their business without thinking about metaphysics at all.

Aviation

The story of a fall, Mircea Eliade observed, could be found in primitive religions all over the world, and so therefore could myths of ascension and flight. These myths expressed a nostalgia for paradise, the desire of man to abolish the weight that confined him to the earth, to convert his body into spirit, and to be free once again to participate in the sacred with the whole of his being.[36] Thus, in its representations of transcendence the Christian symbolic tradition was reprising an older, indeed elemental theme. It also offered something distinctive: the promise of a revelation, and of a reentry into paradise, that would occur (or at least begin) in historical time.[37] Perhaps it is this, along with its sustained allegiance—also unusual among the major religions—to material and spatial conceptions of God, heaven, and an angelic host, that explains why most early experiments in mechanical flight took place in the Christian world.[38] A successful ascent into the skies might be a sign—an actual symbol—of the coming millennium. Yet such experiments may simply have reflected the broader encouragement that Christian doctrine gave to pragmatic technological ventures from the high Middle Ages on. Significantly, it was a Benedictine monk, Eilmer of Malmesbury, who made the first recorded attempt at flight in Christian Europe, in the first decade of the eleventh century, though the story of his descent from a tower of Malmesbury Abbey, widely disseminated in medieval culture, likely chilled in its readers whatever hopes they had harbored for a winged passage into heaven. After gliding for more than a furlong, Eilmer crashed to the ground, breaking both his legs: he "was lame ever after," the chronicler notes.[39]

In medieval dreams of flight, therefore, a religious purpose may have been expressed, but it was clear that faith alone was not enough to keep a man in the air. Hence, in his early painting *The Annunciation* Leonardo da Vinci paid scrupulous attention to the structure of the angelic wings, but later, when he applied his genius to the means by which an actual man might achieve flight, he designed machines. They were machines that also seemed directed toward a secular goal—victory in battle over the enemies of the Italian city-states—rather than toward a communion with God. In the same notebooks that contain his sketches of flying machines, Leonardo drew manifold images of crossbows, catapults, and other military hardware.[40] The exigencies of warfare, indeed, were probably a more immediate stimulus to the uses man subsequently made of the air—for the discharging of munitions and for reconnaissance—than any imperatives of faith.[41] It was deliverance from the specific experience of conflict, rather than from the elemental condition of the Fall, that tended to excite millennial expectations in the early twentieth century, at the opening of the aerial age. The airplane would inaugurate a new era of peace either because it would increase contact and improve communication between the great powers or because its very potential for unleashing destruction simply made warfare unthinkable.[42] Some of these expectations even survived the First World War, most notably in the United States, where they had not been chastened by actual experience of aerial assault.[43] It was there that Antoine de Saint-Exupéry, perhaps the last major writer to make a benediction of flight, to present it as both a source of enlightenment and a catalyst for human concord, found his largest audience.[44]

Even in America, however, those who were most closely engaged in the exacting business of developing new aircraft and testing their capabilities usually refrained from any avowal of a broader sense of mission.[45] Wilbur and Orville Wright were the sons of a bishop in the Church of the United Brethren of Christ, a pietist, evangelical denomination, but they appear not to have inherited their father's fierce religious convictions; nor did their youthful experiences in the church provide clear, decisive stimulus toward experiments in aeronautics.[46] Indeed, for most of their biographers their interest in the problem of human flight seems oddly underdetermined: it is almost as if they woke up one morning and, for want of anything better to do, resolved to invent the airplane.[47] Having succeeded in that goal, the Wrights did not offer their creation freely and immediately to the world but elected instead to keep it under wraps until patents were secured and contracts for sale had been signed.[48] Their commercial ambitions, combined with their habitual terseness whenever they were invited to expound on the wider meaning and significance of what they had achieved, meant that Wilbur and Orville Wright, pioneers of the aerial age, would never become its prophets.[49]

It was only in May 1927 that such a prophet truly arrived, descending from the skies above Paris to land at Le Bourget Field. Compared with those of the

Wright brothers, Charles Lindbergh's motives seemed to be pure: he flew solo across the Atlantic simply to prove that it could be done.[50] It was much easier for commentators to cast his achievement in moral and religious terms. He was described as an angel, a redeemer, who had arrived from the heavens to give new hope to societies that had lost their convictions and squandered their virtue in the debacle of the Great War and the debauches of the Jazz Age.[51] "He was the instrument of a great ideal," wrote Myron T. Herrick, the American ambassador to France, "and one need not be fanatically religious to see in his success the guiding hand of providence."[52] Yet Lindbergh himself, by the time of his flight and for a decade or so thereafter, maintained neither confidence in God nor an attachment to any church.[53] The fate of man, he believed, lay in his own hands: through advances in aviation he might eventually escape the confines of the earth; through advances in science he might evade the inevitability of death.[54] If this was the "winged gospel,"[55] it was a gospel only in form and tone: there was not much of Christian theology abiding in its content. The inspiration for flight may have originated in a dream of restored divinity, but once men had actually ascended into the air, some mistook the means for the end and made of aviation itself a cult.[56] It was a move that left aviation vulnerable to cooption by fascist movements, which sought more deliberately to usurp the prophetic role of the church and to rechannel the passion of the faithful toward the this-world purposes of national renewal.[57] Successes in aviation lent an initial luster of modernity and mastery to Mussolini's regime in Italy.[58] In Leni Riefenstahl's film *Triumph of the Will*, Hitler descends in an airplane from above the clouds—"like a Wagnerian god"—to address the 1934 Nazi Party rally in Nuremberg, symbolically to reclaim Germany from its sixteen years of "suffering" following the war.[59] Charles Lindbergh wrote privately of his admiration for the Führer, and in 1938 he accepted a medal from Air Marshal Hermann Goering in recognition of his services to aviation.[60]

The secular cult of flying did not survive the Second World War. To sacralize the airplane seemed now to be a morally oblivious move, too close to the fascist cultural style and too indifferent to the bloody mess that airborne munitions had made of Europe's cities. Lindbergh paid a visit to postwar Germany and returned transformed: his enchantment with aviation was ended for good, and he had become convinced of man's need for spiritual values and a belief in God.[61] Providential readings of aviation were also changed by the conflict. After Pearl Harbor it was clear even to those Americans who had kept faith in the "winged gospel" that not all airplanes were serving the purposes of Christian peace: it mattered who was flying them, and if it were the wrong side, God would sanction and even assist efforts to send them flaming to the ground.[62] "God is my co-pilot," asserted Colonel Robert Scott of the US Army Air Forces: "My personal ambition is that He permit me to go again into combat against the Jap or the Hun; that He help me just a little to shoot down a hundred Jap ships—even a thousand."[63]

The influence of religious faith on the working lives of postwar aviators was conditioned by increasing professionalization within both the commercial-airline sector and the armed services. The era of mass air transit and the development of high-performance jet aircraft placed a premium on training, procedure, and skill, that is, on reducing the aviator's reliance on courage, the grace of God, or pure, dumb luck. Of course, many seasoned professional airmen still experienced incidents of crisis in the air. For the corps of civilian and military test pilots operating out of Edwards Air Force Base in California, it was precisely their function to press against the limits of their machines. But what generated such crises was characterized in terms of a concept of fate more often than one of divine will, and the test pilots at Edwards rarely regarded fate as a sovereign force.[64] Whatever the emergency, a pilot with the necessary coolness and skill—"the right stuff"—could bring the plane down safely and then join his comrades and some eager young women for an evening of raucous entertainment in the bar of Pancho's Happy Bottom Riding Club, a nearby dude ranch. "Pilot Heaven," Tom Wolfe called such revels.[65] It was a model of paradise very different from the one that had initially inspired man's dream of taking to the skies.

The Rocket Pioneers

A retrospective claim of religious inspiration has also been made for the pioneers of rocketry, those who initially imagined and sought to build the means by which man might venture to the stars. In his seminal history of the space age, Walter McDougall declared that thinking about technology could never be separated "from teleology or eschatology."[66] This was an insight later elaborated by David Noble, for whom the rocket pioneers were either in thrall to a dream of an ascent toward perfection or haunted by the prospect of an end to life on Earth. They invested in their theories and inventions a hope of human immortality: that the species might transcend the death of its home planet.[67]

Some of these men did exhibit a passionate concern with religious or spiritual questions. In Russia, Konstantin Tsiolkovsky, the first theorist to conceive of a multistage, liquid-fueled rocket, was a fervent disciple of the Christian futurist Nikolai Fyodorov, who believed that science would eventually permit the dead to be resurrected and sent out thereafter to populate the cosmos.[68] In Germany, Hermann Oberth developed simultaneously the theoretical principles that would allow a rocket to ascend beyond the earth's atmosphere and a theosophical system that asserted the immortality of the soul and its powers of reincarnation.[69] And in the United States, John Whiteside Parsons invented the first safe and practical solid rocket fuel, while privately nurturing an enthusiasm for black magic and the occult.[70] As Norman Mailer later noted, the outcomes of early experiments in rocketry were so unpredictable—with explosions on stands, sullen refusals to fire, and departures on berserk trajectories—that it would have been understandable if their designers, seeking to petition higher powers, had daubed "the blood of a virgin goat on the orifice of the firing chamber."[71]

Yet not every rocket pioneer thought of his work as an essay into the sacred. Insofar as a common inspiration moved the three fathers of rocket development—Tsiolkovsky, Oberth, and Robert Goddard (the first to actually demonstrate that a rocket could fly)—it came from the novels of Jules Verne rather than from the Bible.[72] Verne himself is best characterized as a kind of Catholic deist, deeply intrigued by the idea of God but unconvinced that he was at work in the world; and Verne was largely uninterested in the figure of Christ.[73] Goddard was not personally religious; his most immediate and consistent motivation was a desire for recognition as the founding genius of rocket science.[74] Frank Malina, who engineered the rockets for which Parsons supplied the fuel and who was subsequently appointed as the first director of the Jet Propulsion Laboratory, had become an agnostic in college after reading Darwin's *Descent of Man*.[75]

Moreover, through the mid-twentieth century, as rocket engineers began to demonstrate the viability of their devices and to court and secure state sponsorship for their development efforts, the more esoteric elements of their culture became increasingly peripheral. After 1917 the Soviet authorities made a hero of Tsiolkovsky but suppressed his religious writings.[76] Oberth elected to maintain a pristine separation between his publications on rockets and space and his reflections on the soul, and what he wrote about the latter rarely attracted many readers.[77] Parsons's deepest immersions in the occult coincided with his retreat from rocket research.[78] In both the Soviet Union and Nazi Germany, certainly, rocket enthusiasts, like aviators, readily declared their allegiance to new ideological cults, and a few became fanatics, but for many it was simply a pragmatic choice. It was only through the state, and specifically the military, that rockets could be brought to a mature stage of development. This was where the resources were; the authorities were closing down amateur rocket operations.[79] The cultic, paranoid character of the regimes for which they worked also confronted the rocketeers with new complexities and hazards. In such an environment the success or failure of their experiments and the speed with which they could convert their devices into weapons available for use against the enemy could be read as tests of ideological commitment. The two preeminent rocket designers of their generation—Sergey Korolev in the Soviet Union and Wernher von Braun in Nazi Germany—were both for a time imprisoned by the secret police, Korolev by the NKVD, von Braun by the Gestapo.[80] It is unlikely that many of the German rocket engineers who moved to the United States after the war felt much nostalgia for the moral-political zeal that had directed rocket development during the Third Reich, even in those years when the US missile programs that employed them seemed without momentum and purpose.

Neither political nor religious millennialism explains Wernher von Braun's personal appetite for building rockets. This enthusiasm, which seems to have evolved out of an adolescent interest in astronomy, long preceded the Nazi rise to power, and it would hardly have been sustained by what was, for the duration of his rocket career in Germany, at best a nominal allegiance to the Lutheran Church.[81] Soon after his arrival in America, von Braun did experience a religious conversion, and from time to time thereafter his writings on the subject of space would present a rather vaporous millennial prospectus. If man was "the master ecological link for all life" in the universe, he asserted in 1960, "then it is profoundly important for religious reasons that he travel to other worlds, other galaxies; for it may be man's destiny to assure immortality not only of his race, but even of the life spark itself."[82] But von Braun's conversion was probably most directly occasioned by a desire to find a new direction for his life after the moral chaos of his service for the Third Reich.[83] If in his postwar career there were moments when he seemed to invest a religious hope in rocket technology, there were also times when he echoed the apprehensions of Charles Lindbergh, declaring religion to be the essential source of the moral law that man required if he was to survive his own scientific and technological inventions.[84] In the nuclear age, the stakes involved in rocket development were even higher than they had been during the war. Described by one biographer as "a twentieth-century Faust," Wernher von Braun had already concluded one bad bargain with the devil; perhaps now he felt a need to have God securely at his side.[85]

Cold War Culture and Politics

Von Braun was not alone. In the early postwar era, from the late forties to the dawn of the sixties, the United States experienced one of its periodic seasons of religious revival.[86] Thousands of new churches were established, with billions of dollars invested in church construction. In the 1950s, church attendance increased by more than 30 percent, compared with a 19 percent growth in the total population. Many of the fastest-growing denominations identified themselves as evangelical or fundamentalist, their recruitment efforts aided in part by the state- and citywide revival campaigns of preachers like Mervin Rosell and Billy Graham.[87] It was Billy Graham, indeed, who, in his successful cultivation of national political leaders, provided reliable evidence for the popular conviction expressed in opinion polls throughout the fifties that the influence of religion on American life was increasing.[88]

The causes of the postwar religious revival were complex. The baby boom, the expansion of the suburbs, and the ambitions of the evangelical movement itself all played a role. It is also difficult to dissociate the cultural salience of religion in this period from the ideological currents of the Cold War, which worked to organize the world into the categories of the god-fearing and the godless. Makers of US foreign policy evoked the concept of a common Christian value system—and a shared commitment to religious freedom—in their efforts to cement the alliance of Western nations against a Soviet Union stringent in its atheism and strongly identified with repression of the church.[89] At home, Christianity, nationalism, and anticommunism melded into a corpus of rituals, rhetorics, and doctrinal codes that at times seemed indistinguishable from an official religion.[90] Congress voted to insert the words *under God* into the Pledge of Allegiance and to adopt "In God We Trust" as the national motto.[91] The armed services expanded the "character guidance" elements of their training programs: aiming explicitly to encourage the development of moral and religious values among their recruits, they assigned the task of instruction to military chaplains.[92] Many local education boards mandated daily prayer and Bible reading in their schools.[93] For those who professed no faith in God, life in the 1950s palpitated with the threat of excommunication from this national church: it was hard for Joseph McCarthy and his confederates to conceive of an atheist whose beliefs had not been cued by a reading of Karl Marx.[94]

Insofar as religious conviction informed American attitudes toward the broader communist menace, it was influential in directing the national response to the specific challenge represented by the Soviet launch of *Sputnik 1*. Americans who held to a traditional faith may have experienced a particular kind of perplexity when they heard the news that a satellite engineered by atheists had made the first claim on God's heavens; and they may also have felt a particular impatience to see that claim contested by devices that bore the imprint of their own Christian culture. However, proposing a policy response on the basis of a defense of the medieval Christian cosmos was hazardous, even for evangelicals. So religious readings of *Sputnik 1* tended to place it instead in the context of a wider, deepening moral crisis: it was important primarily as a symbol of the decadence and mediocrity of American life and thought and of the need, above all else, for further efforts of religious renewal.[95] Certainly, ministers could assert, as Billy Graham did to President Eisenhower, that the safety of the republic required the toughening up of its youth through compulsory physical and scientific training in the schools.[96] They might applaud the launch of an American satellite for the same reasons that they would applaud any other tactical success in the Cold War. But it was also vital, as Graham told Eisenhower, that the American people "be encouraged to look to God who is the source of all our spiritual and moral strength." Indeed, for many evangelicals in the late fifties, the program of modernization that was designed to lay the foundations for the nation's success in the Cold War competition in space entailed precisely the sort of reforms that were most likely to compound the moral crisis they had identified. In particular, the expansion of the federal government's role in education could be expected to degrade the moral autonomy of the individual, the family, and the locality and to obscure the constitutional line of separation between church and state, which allowed God's word to freely work its effects throughout American society.[97] The dawn of the space age was not wholeheartedly welcomed by all Americans of faith.

The primary motivation behind American policymakers' decision to accept the Soviet challenge to a race in space was a concern about prestige.[98] Eisenhower himself doubted that *Sputnik 1* represented much of a national reverse. It did not really matter whether the rest of the world believed that the Soviet Union had overtaken the United States as the most advanced scientific power as long as it was not actually true. He found it difficult, though, to resist the pressure to accelerate development of a space-faring capability.[99] For President Kennedy, investment in such a capability was a national priority only if it would result in a demonstration of US preeminence.[100] He designated a manned lunar landing as the overriding objective of the space program because he had been informed by his advisers that it was something the United States might reasonably expect to accomplish before the Soviet Union.[101]

When justifying the space enterprise to a public audience, however, policymakers did not always speak first, or most of all, about the calculus of international prestige.[102] Such discretion made sense: we are not always most impressed by the things that we know are done to impress us. Better to pitch the program as if it had been inspired by the passion for exploration and the thirst for knowledge and let the world infer a lesson about the power and self-confidence of the nation that was ready to direct massive resources to the pursuit of wonder and enlightenment.[103] It was here, in the articulation of the existential rewards that might accrue from man's voyages into space, that policymakers might have made reference to religious hopes for the venture. Such references were actually uncommon or else enveloped within language that could as easily be read as an anticipation of secular scientific returns.[104] "With its hint of the discoveries of fundamental truths concerning man, the earth, the solar system, and the universe," noted one official policy document in 1958, "space exploration has an appeal to deep insights within man which transcend his earthbound concerns."[105] It was difficult to declare a more precise religious inspiration for American adventures in space without weakening their appeal to the large, non-Christian sectors of their principal market: international opinion. Individual astronauts could, within reason, say what they liked about God or say nothing at all, for that—as Kennedy observed following John Glenn's orbital flight in *Friendship 7*—reflected the religious liberty in which they lived and worked. Kennedy contrasted Glenn's statement that he had not prayed during his flight to the rote hosannas to the communist system sounded by Soviet cosmonauts.[106] According to this same logic, those in executive authority had to be more circumspect still, for they might find their personal avowals of religious hope regarded as a prescription for a Christian crusade in space.

The NASA Leadership

Insofar as it is possible to discern the religious convictions of the early leaders of the space program, it appears that only a few considered their work in this field to be a commission from God and a fulfillment of his providential purpose. There were certainly senior managers in the program who can be classed as devout: Hugh Dryden, NASA's deputy administrator from 1958 to 1965; Major General John B. Medaris, who headed the army's missile-development program from 1955 to 1960; Eugene Kranz, flight director at the Manned Spacecraft Center (MSC) in Houston during the Gemini and Apollo missions; and, of course, Wernher von Braun, who ran the Marshall Space Flight Center in Huntsville, Alabama, from 1960 to 1970. Yet it was really only von Braun who folded space exploration into a millennial theme, and he did not do so often.[107] More usually, he would argue modestly that religious faith was compatible with modern understandings of the natural world or, in a rather different vein, that mankind needed religious values in order to keep modern science and technology in check.[108]

Like von Braun in his more pessimistic moods, Dryden, a lay Methodist preacher, was inclined to doubt that man's prodigies of material invention were really leading him to God. Scientific and technological advances, he believed, were implicated in the spiritual impoverishment of modern civilization; they were the product of a view of man as "a physio-chemical system, a machine, rather than a soul to be saved or lost." There were only a few "research workers" who questioned this dismal materialism, who sought to use their talents "unselfishly to add to the spiritual development of mankind."[109] Medaris claimed a divine guidance for his decisions, but he seems to have been mostly concerned with preserving the army's role in rocket development, not with the goal of space exploration itself. Moreover, he judged himself to have become a true convert to Christ only after he left the army, following a cancer diagnosis in 1964.[110] And for Eugene Kranz, entering the Manned Spacecraft Center on the morning of the first moon landing, there was "my usual vague feeling that somehow my entire life has been shaped by a power greater than me to bring me to this place, at this time." Still, as he also recalled, the most immediate stimulus toward a career in aerospace engineering had been his youthful fascination with rockets and a personal love of flying.[111]

These individuals aside, the men who inhabited the upper levels of NASA management were not, on the whole, conspicuous in their piety. There is little evidence of an animating religious faith in the diaries of T. Keith Glennan, the agency's first administrator, or in biographies of James Webb, who presided over the agency's rapid expansion in the wake of Kennedy's decision to take America to the moon.[112] Indeed Webb was not initially even very interested in spaceflight; though he quickly warmed to his task, it was because the program offered a perfect laboratory for his theories about administrative leadership.[113] "I've always been religious," asserted Robert Gilruth, the first director of the Manned Spacecraft Center, but he did not attend church regularly. Gilruth's own account of his pivotal role in the development of the nation's manned space capabilities provides no examples of the interventions of faith. His job as an employee of a federal agency was simply to do whatever the country required at the time; in the sixties, given his expertise in aerospace-engineering management, this just so happened to involve working out a way to fly to the moon.[114] Christopher Kraft, director of flight operations at the Manned Spacecraft Center, served as a lay reader at his local Episcopal church, but he admitted that these two offices seemed to exist in separate worlds: "I tried teaching a Bible class, but I lacked the fundamentalist verve and drove people away when I tried too hard to relate the early church to more modern interpretations. It was hard not to be modern when I spent my working days sending man into space."[115]

What issued from the miscellaneous private religious dispositions of the NASA leadership was not a Christianized space program. For the senior managers who shaped the culture of the agency in its first decade, the fields of faith and work were not completely segregated, nor did one make substantial claims on the other. This was a pattern rarely articulated in policy terms but discernible nonetheless. NASA adopted a liberal leave policy for staff wishing to observe religious holidays; and it allowed even those astronauts involved in the urgent task of postflight debriefing to attend church services on these holidays if they so desired.[116] Astronauts were permitted to carry small religious items into space as part of their personal-preference kits.[117] The agency's managers were also alert to the likelihood, given the connotations of an ascent into the skies and given its dangers too, that some astronauts might want to make a testament of their religious beliefs and that the public might want to hear them. After Gordon Cooper recorded a private prayer during his orbital flight aboard *Faith 7*, Robert Gilruth and Deke Slayton, head of the Astronaut Office, suggested that he use it to close the speech he had been asked to deliver before a joint session of Congress.[118] Gilruth and George Mueller, head of the Office of Manned Space Flight, also approved the *Apollo 8* crew's plan to read from the book of Genesis during a broadcast from lunar orbit on Christmas Eve, 1968.[119]

The ambition of NASA managers, as well as their political masters in the White House, was that the whole world would feel some sense of common ownership of the endeavors of the United States in space. That ambition would not be served by evoking too precisely the aid and inspiration of a Christian God. By the mid-sixties the constitutional environment was changing back at home, as the 1962 and 1963 Supreme Court decisions prohibiting school prayer reminded public officials that any effort to direct religious activity or expression under the auspices of the state could conceivably result in a legal challenge. Certainly, "the moral and spiritual well-being" of the country was "an important consideration," wrote one NASA public-information officer in 1966, in response to a church minister who had called for the "official inclusion of a spiritual aspect" in the national space program. Nevertheless, the "primary purpose" of the program was scientific "and must continue to remain so as long as it is financed by funds obtained by taxation for a public purpose, as provided for by the Constitution of the United States."[120] Thomas Paine, who succeeded James Webb as NASA administrator, could characterize the program's successes as triumphs of "the squares—meaning the guys with crewcuts and slide rules who read the Bible and get things done"; he could tell the National Conference of Christians and Jews that the Bible was "the best operating manual ever written."[121] But Paine was not notably devout himself, and this was about as far as he, as well as most of those around him in NASA leadership circles, was willing to go in making a common cause of spaceflight and religion. It was a point noted by one NASA scientist, an evangelical, on the eve of the first moon landing: "If the space program can be faulted for anything," Rodney W. Johnson told *Christianity Today*, "it is that it has ignored man's spiritual yearnings."[122]

NASA's Institutional Culture

Norman Mailer, who had prided himself on his capacity to divine the essence of any event, was unusually perplexed by the mission to land a man on the moon. As an institution NASA was inscrutable. What was one to make of its apparently divided nature, that it was engaged in an enterprise that spoke of a grand metaphysical ambition yet almost autistic in its focus on the technical tasks at hand and its inability to talk about what it was doing, and why, in language that was not either banal or diseased with jargon? Perhaps, Mailer reflected, technological man, working through NASA, had taken up the quest of traveling to the moon because, however dimly, he perceived it as his last best hope for escaping his own instrumentalism and participating once again in ontological mystery. Alternatively, the venture might be a sign that technological man had gone insane.[123]

It is probably true that visionaries and prophets were not much in evidence in the middle ranks of the space program. M. G. Lord, whose father worked as a contractor at NASA's Jet Propulsion Laboratory in California, depicts him as the model of the Cold War rocket engineer "who embraced the values of his profession: work over family, masculine over feminine, repression over emotion."[124] Apollo-era NASA engineers tended to match the broader profile of their profession presented in the sociological studies of the sixties. They had been drawn to the field by a pragmatic judgment that it would lead to a stable career; before entering the workplace, they had received a narrow technical education, with little exposure to the liberal arts; and once employed, they tended to become more specialized still, rarely involved in basic research or encouraged to develop holistic perspectives on the projects on which they worked.[125] Unlike their peers in other institutions, however, the NASA engineers rolled out their projects before the gaze of the entire nation. Any malfunction would be painfully public; and on the manned space program—as the 1967 *Apollo 1* disaster proved—it might also entail the death of an astronaut crew. It is hardly surprising that these engineers were intensely focused on making sure that their machines performed properly. Indeed, if there was one guiding principle behind the development of NASA's manned space capability, it was that as little as possible should be left to the work of prayer; hence the many redundant systems integrated in the rocket and spacecraft designs, the hundreds of pages of mission rules, and the long weeks that each crew spent in simulators prior to flight.

Yet a kind of idealism could be detected in the diligence and care that NASA technicians directed toward their work.[126] There was something in such professionalism that spoke of a dissolution of self in performance of a sacred trust. To Mailer, the agency's personnel behaved like "true Christians"; to Oriana Fallaci, they "constituted a religious sect: ready to sacrifice and deaf to irony."[127] In aspects of NASA's cultural style, aesthetics of futurity intermingled with evocations of the medieval church. At Cape Canaveral, on the nights just before a mission the rocket on its launch pad was bathed in white light, resembling a holy shrine.[128] The astronauts were confined to what Michael Collins called the "monastic world of crew quarters," men seeking seclusion in pursuit of the ideal.[129] In Houston, similarly, the doors of the mission-control room were locked just prior to the launch. Like a priest donning vestments, Flight Director Kranz put on an elaborate waistcoat specially made for each mission by his wife.[130] The countdown proceeded like a liturgy, complete with its own ritual language. After the launch, a vigil was maintained through night and day until the crew were safely returned. And after their return, the debriefings began; "almost like confessing my sins to a priest," Kranz observed.[131] Perhaps these gestures and routines were evidence of a God-haunted world. Although each may have been explicable in instrumental terms, they drew their solemnity and their power to direct and discipline thought from analogues in sacramental practice. But in such analogies was the substance of religion transferred, or simply its form?

The answer may depend in part on where we choose to look. The three NASA centers most actively involved in the Apollo program—the Marshall Space Flight Center in Huntsville, the Manned Spacecraft Center in Houston, and the Kennedy Space Center at Cape Canaveral—each manifested its own distinct cultural style, influenced by the religious and social norms of the community in which it was located and by the content of its NASA missions and the work routines that attended them. The massive expansion of the agency's operations through the early to mid-1960s contributed in these three communities to high levels of in-migration from elsewhere in the country and thus to marked increases in the local population. They became hubs of employment opportunity, not just with NASA but also with its contractors and subcontractors and with the hundreds of companies in the surrounding areas who benefited from the space program's economic multiplier effects.[132] Some of the same localized population impacts were stimulated by federal largesse in other locations too, most obviously across the "gunbelt," in the nodes of postwar military-industrial production.[133] Around the centers of the Apollo program, though, these changes occurred more quickly than anywhere else. In 1970 the population of Brevard County, which included the Kennedy Space Center, totaled more than 230,000, compared with 23,653 twenty years earlier.[134] For a time, it was the fastest-growing county in the nation.[135] Later, as the Apollo program was closed down and NASA was obliged to reduce its workforce, all three of these communities confronted the challenge of economic diversification, of generating alternative sources of employment sufficient to meet the needs of a newly enlarged population.[136]

In Huntsville, Brevard County, and the suburbs clustered around the MSC, southeast of Houston, few social institutions did not feel the effects of the local space booms. Churches were among the institutions most strongly marked by the in-migration. Once again, some of the changes they experienced were not confined to communities with NASA installations in their midst. Across much of the country, in the wake of a "continental movement" of population from the Northeast and the Midwest to the South and Southwest and a "local movement" from urban centers to their peripheries, religious institutions assumed the task of ensuring that families taking up residence in new and unfamiliar neighborhoods had convenient access to places of worship; otherwise, over the course of these migrations millions might be lost to God for simple want of a nearby church. Denominational authorities actively sponsored the establishment of new suburban churches, while existing congregations expanded their operations, increasing the size of their sanctuaries and their Sunday-school facilities, and in many instances seeded satellite congregations of their own.[137] The increased church attendance in fifties America can probably be taken as evidence of the initial success of such efforts. But the crowded pews of the suburban churches did not convince everyone that across the crab-grass frontier a true Christian faith was in full bloom. This was the era of the suburban jeremiad, which cautioned that the mission of the suburban church was in danger of becoming degraded. It was not communion with God that suburbanites sought when they attended Sunday services, nor was it devotional commitment that motivated them to participate in other parish activities. Rather, what they wanted from the church, and what the church worked to provide, was the opportunity to meet one another. The role of the church in suburban America, according to these jeremiads, had devolved to that of a social center, distinct from the Rotary Club and legion hall only in the nature of the reward it offered for nominal involvement: an assurance of cheap grace.[138]

In 1940 Huntsville was a relatively small southern town with a metropolitan population of about thirteen thousand and a cultural climate consistent with its location in the heart of the Bible Belt.[139] The establishment of the Huntsville Arsenal and the Redstone Ordnance Plant on the eve of America's entry into the Second World War brought jobs and economic prosperity, at least for the duration of the conflict, but also the apprehension that Huntsville, like other boom towns, would succumb to moral permissiveness. Hence, after a campaign led by religious leaders, Huntsville voted to go dry.[140] It was into this environment that Wernher von Braun and his team of German rocket engineers were inserted in 1950, following their recruitment by the US military and a prolonged period of purgatory in the Texas desert tinkering somewhat aimlessly with the V-2 missiles they had developed during the Third Reich.[141] With the escalation of the Cold War, the two ordnance installations in Huntsville were reactivated and the Germans set to work on a new ballistic missile.[142] To the town itself, they brought a measure of cosmopolitanism: their interest in literature and music was reflected in the construction of a new public library and the creation of the Huntsville Symphony Orchestra.[143] Over the course of the next twenty years the Germans and those who followed them to Huntsville to take up positions in the missile-space economy did much to promote improvements in the city's education system, raising standards in the public schools and establishing a local campus of the University of Alabama.[144]

Though they proved to be agents of change, the Germans also found aspects of Huntsville's existing civic culture congenial. Like most citizens of the town, they took religion seriously, moving to organize their own Lutheran church almost immediately after their arrival. In this endeavor they were assisted by local Episcopalians.[145] Moreover, there was probably an affinity between the values Huntsville's civic leadership professed and the conservatism that characterized the Germans' approach to rocket engineering. Both preferred that innovation, whether social or technical, proceed only in bite-sized increments.[146] During the 1960s a spirit of religious revival seems to have blown though sections of Huntsville society, and it did not leave the Marshall Space Flight Center untouched.[147] In 1962 Billy Graham held a rally at the Redstone Arsenal; von Braun spoke often about his own faith; and William R. Lucas, who became director of the center in 1974, was a prominent Southern Baptist layman. Religious principles, Lucas asserted, had not motivated him in his work as a rocket scientist, but now that he was a NASA manager in an era of retrenchment, they guided his desire to make decisions that were fair. "The vast majority of people at Marshall," he later recalled, "were Christian people."[148]

The MSC in Houston was tasked most directly with landing an astronaut on the moon by the deadline set by John F. Kennedy in 1961—"before this decade is out." For workers at the center, then, the clock was always ticking. The origins of the MSC lay in the National Advisory Committee on Aeronautics (NACA) laboratory in Langley, Virginia, an institution strongly oriented toward experimental research and also fiercely protective of its own independence and that of its engineers.[149] Out of this inheritance, along with the implacable pressure of time, evolved a center culture characterized by an emphasis on getting the job done even if that involved bypassing bureaucratic hierarchies and making sudden improvisations around established development schedules. The proposal to send *Apollo 8* around the moon without a lunar module had been conceived in Houston; it was the sort of play that distinguished the MSC from Marshall, with its reputation for technological conservatism.[150] Yet, if the institutional style of the center admitted extemporization, it was intolerant of gaiety. Its employees were like the Puritans, making a claim on the heavens through self-discipline and hard work, fearful of the meaning of laughter. "Away from Houston," one official told Oriana Fallaci, "we can throw our hats in the air if we meet a friend. In Houston we stiffen up like boys in school. We watch each other, we spy on each other, and we're in terror of being thrown out."[151] When members of the press, watching in the MSC movie theater, started to giggle as Neil Armstrong maneuvered slowly and awkwardly through the lunar module hatch toward his first steps on the moon, Norman Mailer noted the incongruity: "Sanctimony at NASA was a tight seal. A new church, it had been born as a high church. No one took liberties."[152]

The MSC complex was constructed and staffed in just over two years, and once it was established, its employees became habituated to often punishing work routines.[153] For many of their families, then, a double dislocation was involved: they moved to the Houston suburbs, and thereafter husbands and fathers absented themselves from the household almost entirely for long stretches each week. Local churches came to play an important role in the lives of these families. They served as an entry point into suburban social networks, they provided a connection to cultural tradition and an external source of moral support amid the grind of modernity, and they offered an opportunity every Sunday for families to go somewhere together, as if to confirm to themselves and to the world that they continued to exist more than simply in name. The MSC may have been one kind of church, but it did not dissolve the need for another. Indeed, it may have made that need rather more acute. The ministers of churches in suburban communities like Webster, League City, Nassau Bay, and La Porte frequently became immersed in the hustle and bustle that accompanied the Apollo missions. They held special services before the launches and sometimes attended the launches themselves; they spoke to the news media about astronauts who were members of their congregations; they visited the astronauts' families.[154] One local minister organized a prayer league, largely funded by donations from workers at the MSC, with the aim of encouraging prayers for the success of the missions and the safe return of their crews.[155] When Buzz Aldrin celebrated communion after *Apollo 11* landed on the moon, he used bread and a wine chalice provided by Dean Woodruff of Webster Presbyterian.[156] Perhaps all this was evidence of a seamless community of faith, embracing, as in Huntsville, the center and its area churches. But the relationship does not seem to have been as easy as that. The center and the churches were codependent, and the traffic between them steady, precisely because they constituted different worlds, each offering something unique but lacking something essential. It was in space, not in the suburban church, that technological man looked for his challenge. Success in meeting that challenge could not compensate fully for its costs—to family most of all. It was in the hope of mitigating these costs that, in Houston at least, technological man turned to the church.

Before 1950, when the US Air Force established a missile-testing facility at Cape Canaveral, Brevard County had been a quiet, lightly populated place. Its residents made their living from citrus groves, fishing, and a steady trickle of tourists who sought the warmth of the Florida sun but wanted to avoid the state's major resorts.[157] The missile age changed all that, but its impacts on the local culture came themselves to be overwritten with the arrival of NASA, the declaration of national designs on the moon, and the subsequent construction of the Kennedy Space Center. In twenty years the county experienced a tenfold increase in population. Its churches responded by rescaling their ministries: they sought to reach out to the new arrivals, encouraging them to become members; they enlarged their sanctuaries and educational facilities to accommodate rising numbers of congregants and Sunday school enrollments. The First Baptist Church in Cocoa, the town on the mainland located closest to Cape Canaveral, sponsored missions throughout the conurbation that was developing around it as a result of missile-space activity.[158] In time, all of these missions—on the west side of Cocoa itself, on Merritt Island, and in Rockledge, Frontenac, and Cocoa Beach—matured and became independent congregations. Riverside Presbyterian in Cocoa Beach, meanwhile, owed its origins to a decision of the national Presbyterian council to establish a church in the town. Until 1957, when a dedicated church building was constructed, the congregation met for its services in the private homes of its members and a variety of other locations, including an Episcopal church, a room above a restaurant, and a bowling alley.[159] Guenter Wendt, in charge of the launch pad at Kennedy, belonged to the church.[160] In late 1961 and early 1962 Riverside Presbyterian attained a measure of national prominence, and experienced something of a media scrum, when John Glenn participated in its Sunday devotions as he awaited his flight aboard *Friendship 7*.[161]

But religious institutions in Brevard County lacked the cultural salience of their counterparts in Huntsville and Houston. The space-age in-migration may have brought more people to their services, but few of the new congregants fully committed themselves to the broader missions of the churches they attended. Local clergymen reported that "there seemed to be no way in which the newcomers could be made an integral part of the church community."[162] In 1965, according to most quantitative indices, including attendance and income, the First Baptist Church in Cocoa was thriving, but its members were chastised by one of the ministers for their "laxness and complacency." The church was distinct from other kinds of organization, he told them: it was a new creation, and its purpose was to evangelize.[163] But both participation in outreach and the success of the outreach effort were likely to be limited in a community where many inhabitants understood that their future lay elsewhere. NASA relied heavily on contractors, and those who worked for such contractors knew that they and their families were fated to lead an itinerant life, to follow contracts around the country, to set up house for their duration and then to move on. This sense of social impermanence, however, was particularly acute in Brevard County, for it was there, in the early to mid-1960s, that the Apollo launch complex at the Kennedy Space Center—one of the largest building projects in the nation's history—was being constructed.[164] Though the complex, once operational, would generate additional employment opportunities (at least for the length of the Apollo program), many of the engineers and workers who established residence in the county while they helped to build the complex expected to leave once that task was complete. It may be no coincidence, then, that First Baptist started to experience a decline in actual attendance at its services in 1967, the year that Apollo launches began at the Cape.[165]

The churches clustered around the Kennedy Space Center were forced to contend not just with the rootlessness of their parishioners but with a local cultural milieu on which Christian moral teaching seemed to exercise no traction at all. As in Houston, the pressures that accompanied employment in the space program took their toll on marriages and family life. In Brevard County, however, their effects were compounded by the abundance of bars and motels and by the presence of a fluctuating colony of young, unattached women drawn to the area by its beaches, the spectacle of the rocket launches, and the association with national celebrity in the person of the astronauts.[166] According to one minister, infidelity was so common that it had become a community joke.[167] As the Reverend Robert L. Lowry, pastor of Riverside Presbyterian from 1969 to 1975, later recalled, "We were enmeshed in a culture that gave little strength to the values we espouse. Our community was known throughout the fifty states as a 'fast' place."[168] Riverside itself offered no secure sanctuary against the carnival outside. Lowry's predecessor had been forced from the church after he and his wife divorced, and in the wake of charges of "inappropriate behavior."[169] It was partly the hope of changing Cape Canaveral's sex-and-booze image—along with the prospect of a capital investment that might mitigate the impact of reductions in NASA's budget—that motivated local civic leaders in late 1970 to welcome the fundamentalist preacher Dr. Carl McIntire's plans to establish a "Reformation Freedom Center" in the town. The proposed development was to include a Jerusalem theme park, a Christian college, a two-hundred-room conference hotel, and a retirement community.[170] These leaders came to choke on their initial enthusiasm. McIntire's ambitions exceeded his ability to pay for them, and the elements of the center that did open to the public found that the public did not come.[171] The local chamber of commerce complained that McIntire had actually driven convention business away with his ban on the sale of alcohol in the Freedom Center Hotel, which was Cape Canaveral's largest conference facility.[172] It was the final twist in a poignant tale: in Brevard County, religion was always the stepchild of the space age, never truly at home and, though occasionally embraced, never the object of unconditional love.

The Astronauts

In 1968 the political scientist Victor C. Ferkiss attended the annual convention banquet of the American Society for Public Administration. The speaker at the banquet was an astronaut. Here, Ferkiss observed, was "one of the few members of our species who had done what men throughout human history have dreamed of doing." The astronaut, whom Ferkiss did not name, had viewed his home planet from afar; he had walked weightless in space; he had crossed the "cosmic boundary" that had confined the fate of mankind to that of the mortal earth. Strikingly, in his speech there was no trace of interest in these sublimities. His account of his experiences in space was sincere but clinical. Indeed, the only false notes were sounded in exactly those passages "when he spoke of wonder felt or beauty observed." Ferkiss concluded that it was not from such sources that the astronaut derived his satisfaction but rather from his successful synthesis with technology. He "seemed so obviously at peace with himself and his world, so perfectly adjusted to the machines and the organization around him, knowing them and their capacities as well as he knew himself, finding in that knowledge peace and freedom."[173]

Ferkiss was not the only student of the space program to identify a discrepancy between the endless horizons that it had apparently opened up for human exploration and the poverty of vision that seemed to afflict the men commissioned to light the way. Norman Mailer offered this characterization of the astronaut: "powerful, expert, philosophically naïve, jargon-ridden, and resolutely divorced from any language with grandeur to match the proportions of his endeavour."[174] Oriana Fallaci interviewed six astronauts in Houston and wondered at their sameness. This, she decided, was the price they had paid for the hope of going to the moon, and the price was too high, for what was the point of human exploration if it cost the explorers their humanity? "So wake up, stop being so rational, obedient, wrinkled," she implored them. "Laugh, cry, make mistakes, disobey!"[175] It was the space program's brand of "tough, understated, essentially anonymous" astronauts, asserted one White House aide in 1973, that was partly to blame for the decline in popular interest in its activities and public support for its expenditures.[176] Ferkiss's astronaut "could just as well have been a rising junior executive in any large American corporation."[177] What the space program needed instead, perhaps, was shamans.

Certainly, not many astronauts in the Mercury-Apollo era responded freely and comfortably to inquiries about the spiritual and philosophical meaning of their voyages in space. Their reticence may be attributed to the sensed deprivations that attended their college education in engineering: in such programs who would have taught them to speak of such things? It was consistent, too, with the laconic style of the test-pilot community, which itself expressed an intuition that philosophy was not efficient. To dally with existential questions was to take one's mind off the job, which could prove lethal. The point seemed to have been reaffirmed in the early years of the space program by the fate of Scott Carpenter, who was probably the most cosmopolitan of the Mercury Seven. Carpenter was judged to have become so distracted by the aesthetic spectacle of space during his flight aboard *Aurora 7* that the mission nearly ended in disaster. He never flew again.[178] Newsmen, Michael Collins observed, would probably have preferred an Apollo crew comprising a philosopher, a priest, and a poet to one manned by three tight-lipped test pilots, and they would have looked forward especially to the press conference that would follow the crew's return from the moon. Sadly, Collins continued, "they wouldn't get them back to have the press conference, in all likelihood, because this trio would probably emote all the way back into the atmosphere and forget to push in the circuit breaker which enabled the parachutes to open."[179] The currency of such perceptions in and around the NASA Astronaut Office may well explain why it took so long for those members of the astronaut corps who conspicuously maintained an interest in matters other than just flying—Bill Anders, Alan Bean, Walt Cunningham, and Rusty Schweickart—to be entrusted with prime crew assignments.[180]

So it was that when the Mercury astronauts were first introduced to the press in April 1959 and each was asked whether he held a religious faith that would sustain him through the challenges ahead, most of the seven dutifully identified the denomination to which they belonged. But it was evident from their answers that they did not look to religion for their motivation or for an assurance of success. Scott Carpenter replied that although he had religious faith, "I don't call on it particularly associated with this project." Deke Slayton took the same view: "I don't feel that any particular extra faith is called for in this program over what we normally have." "I think I should like to dwell more on the faith in what we have called the machine age," declared Wally Schirra. "All of us have had faith in mechanical objects."[181] It came as a surprise to these men that religion might be thought relevant to their work.[182] As their enthusiastic embrace of the carnal pleasures readily available to an astronaut in the bars and motels of Cocoa Beach indicates, it was not even very influential in their private lives.[183] The exception was John Glenn, for whom church, work, and family seemed genuinely to exist in a continuous field. At that first Mercury press conference, Glenn explained his belief that if a man used his God-given talents properly, if he lived "the kind of life we should live," there was "a power greater than any of us" that would place opportunities in his way and ensure that he was "taken care of."[184] "John tries to behave," commented one of his friends, "as if every impressionable youngster in the country were watching him every moment of the day."[185] Still, Glenn's religious convictions were not so highly evolved that they explicitly directed his labors toward exploration of the heavens. He joined NASA with an interest in space that was both rather recent in vintage and largely motivated by secular concerns.[186] Not the least of these was the recognition that, having never received a college degree, he was unlikely to advance much further in the military hierarchy.[187] Hence, when NASA came calling, Glenn was most certainly in the market for a change of career.

Periodically through the mid-sixties, as Mercury led to Gemini and Gemini led to Apollo, and as its spacecraft expanded to accommodate two men and then three, NASA issued recruitment calls for additional astronauts. The men who entered the astronaut corps as a result of these calls were more diverse than the Mercury Seven in their background, beliefs, and motivations. For the third call, in 1963, the agency dropped the requirement that candidates had to have served as test pilots. Though some of NASA's new recruits still counted flying as their original inspiration, embracing the space program much as they had any other kind of equipment upgrade, they now found themselves in the company of men who variously expressed a fascination specifically with rockets, a commitment to science, and an interest in the ends as well as the means of exploration.[188]

By the mid-sixties the astronaut corps, though still hardly representative of the nation as a whole, had certainly become a somewhat broader church. This was evident, appropriately enough, in the religious orientations of its members. In William Anders and Jim McDivitt the corps could now claim two devout Catholics.[189] Frank Borman, Edward White, and David Scott, meanwhile, all seem to have adhered to a strain of tough, unfussy mainline Protestantism readily compatible with the ethos of military service: God went with country, and so military families went to church. Borman was a lay reader at St. Christopher's Episcopal Church in League City, and his wife Susan taught Sunday school there; Scott was a vestryman at the same church.[190] There were those within the corps, however, who maintained their distance from organized religion. Mike Collins, for whom wry circumspection was virtually a default setting, generally did not accompany his wife and children to their Catholic church.[191] Neil Armstrong, asserts his biographer, is most accurately described as a deist.[192] Edgar Mitchell had been raised as a Southern Baptist, but only the barest vestiges of that faith—"some hidden fear of God"—survived by the time he became an astronaut.[193] Rusty Schweickart was not religious, and Walt Cunningham, by his own account, was "at best an agnostic in most people's eyes."[194]

And then there was Buzz Aldrin. To scan the surface material of Aldrin's life and career as he entered the space program and eventually took his place on the *Apollo 11* crew would probably be to conclude that here was another John Glenn, the model of a dedicated Christian astronaut. Aldrin was an elder of Webster Presbyterian, where Glenn had also worshipped; he taught Sunday school at the church, as did his wife Joan.[195] He donated 130 copies of a condensed version of the Bible, entitled *Good News for the Modern Man*, to the church in memory of his mother and presented another copy to Guenter Wendt on the Apollo launch pad just prior to his departure for the moon.[196] Aldrin marked his arrival on the moon by serving himself communion, "symbolizing the thought that God was revealing himself there too, as man reached out into the universe."[197] Finally, in a television transmission as the crew was headed back to earth, Aldrin reflected on the "symbolic aspects" of the *Apollo 11* mission and quoted from Psalm 8: "When I consider the heavens, the work of Thy fingers, the moon and the stars which Thou hast ordained, what is man that Thou art mindful of him?"[198]

Something, however, did not quite fit. Unlike Glenn, Aldrin had received an advanced technological education, and there were elements of his personality that conformed more to common typologies of the engineer than to the immaculate incarnation of midcentury Christian masculinity that Glenn had offered to America in the early 1960s. Aldrin immersed himself in the theoretical and technical content of his work and achieved a measure of genuine mastery. But when he sought to engage in bureaucratic politics or to participate in the social rounds of the astronaut life, he often appeared maladroit, tone-deaf, and ill at ease.[199] Aldrin's involvement with Webster Presbyterian presented him, perhaps, with a rare chance for social integration, for performing a role within the community for which the cues were easily readable and precedents well-established. His faith, Norman Mailer asserted, "was as predictable as a flow-chart he had designed himself"; its purpose was "to restore emotional depletions."[200]

Certainly, to read Aldrin's memoirs is to conclude that it was not religious conviction that set him on a career path to the moon. It is more likely that he was driven by the expectations of his father, a former air force colonel, who had known Orville Wright, Robert Goddard, and Charles Lindbergh and who had taught the young LeRoy Grumman, subsequently founder of the corporation that would make the Apollo lunar module.[201] The elder Aldrin was an auxiliary to a tradition of aeronautical innovation that, by the time it took his son into space, had attained functional autonomy from whatever religious aspirations had lain at its source. Significantly, when Aldrin himself was asked about his motivations, he was unable to provide an answer that connected action to value: "There's a treadmill that's going, and if you make it, you're on it." He continued, "Why do you do anything? Maybe because you were selected to do it."[202] Whatever his attendance at Webster Presbyterian had offered him, it was not a compelling Christian rationale for traveling to the moon. Aldrin used the rite of communion to distinguish *Apollo 11* from other, purely secular displays of technological prowess, but the rite alone could not define in what way the mission was actually sacred. The technocrat in Aldrin may have intuited that an effort to bend Christian theology to the lunar venture would itself be a kind of technocratic adjustment. Hence, perhaps, his request to his pastor, the Reverend Dean Woodruff, "to come up with a symbol which meant a little bit more," which "transcended modern times."[203] Hence also, perhaps, Woodruff's turn to ancient myth in his search for the elemental meaning of Aldrin's ascent to the moon.[204]

The space program of the Mercury-Apollo era would not have taken the form that it did, and could not have achieved its principal goals as speedily as it did, without the labor and commitment of religious Americans. The program was no bastion of secularism: elements of NASA's institutional style and at least some of its activities spoke to an open concourse between place of worship and place of work. But the thesis that the program was conceived and brought to term in anticipation of a Christian millennium, enforced by analogous movements of ascent and promises of immortality, is not persuasive. Indeed, religious faith was rarely an inspiration for the project of spaceflight; it was salient mostly as a source of validation. In the context of the religious Cold War, and in the suburban communities where churches served as the centers of local social routines, it mattered that the way a man earned his living was broadly compatible with Christian doctrine and conscience. Christian doctrine and conscience, however, generally did not determine the specific choice of a career in spaceflight in preference to any other vocational field. Nor did they exert much influence on how, day to day, that man went about his work. Through the maturation of the enterprises out of which spaceflight evolved—technology and aviation—and into the actual age of rocketry and space travel, the imperatives of religious faith seem to have progressively lost their force. Those engaged in these enterprises were increasingly moved by self-interest, or by the secular ideologies of nation and state, or by a loyalty that seemed both derived from and confined to the activity itself—to engineering, to flying, to sending a rocket to the stars. In an organization like NASA it could be hard to make the cognitive journey from what was often a highly specialized work role to any kind of cosmic view. Exacting institutional codes and quality-assurance practices articulated a sense of the distance that lay between the social and familial realm, where church teachings could plausibly make a difference, and the intricate operational spaces of a modern technocracy, especially one that, in seeking to place a man on the moon, labored within conspicuously unforgiving margins for error. In the space-age communities around Brevard County, indeed, it was not clear that the churches made much difference even in the area of private morality.

Not secularism, then, but secularity. The space program, for all of the Christians in its midst, for all of its evocations of transcendence, was a product primarily of profane, sometimes prosaic, ambition. For the most part, indeed, it serves as an object study in differentiation: religious values and symbols often did make the commute from suburban altar to NASA space center, but they were usually weakened by the journey, to the point that they exerted no autonomous authority over the substance and direction of space policy. NASA managers made little effort to explicitly sacralize the agency's activities. The word of God was, with rare exception, conspicuously absent from the radio traffic between NASA spacecraft in the heavens and its installations on Earth. It was mostly left to those outside the program, if they so wished, to reflect on the relation between exploration of the cosmos and cosmological tradition, to judge whether the space age would be friendly to faith.

CHAPTER TWO

Signals of Transcendence

The Rise and Fall of Space-Age Theology

"Without risk, no faith," declared the Danish philosopher and theologian Søren Kierkegaard in the 1840s. "If I am able to apprehend God objectively, I do not have faith; but because I cannot do this, I must have faith. If I want to keep myself in faith, I must continuously see to it that I hold fast the objective uncertainty, see to it that in the objective uncertainty I am 'out on 70,000 fathoms of water' and still have faith."[1] Replace the bottomless ocean with the depths of the universe, speed on more than a century, and this text could be read as a better than adequate summary of the response religious thinkers presented to the theological challenges of the space age. There was nothing in this era of satellites, rockets, and interplanetary travel that fundamentally reformed man's condition of "objective uncertainty," as Kierkegaard described it. The same existential question remained: how can man know where he stands in relation to God? The only difference now was that it was asked in full knowledge of the dizzying dimensions of the cosmos. The theater had grown in size, the players had more impressive props, but the essential drama was unchanged, and so too was its message: it was uncertainty that made faith possible as well as necessary. In the words of Martin J. Heinecken, a Lutheran theologian and Kierkegaard scholar who authored the most complete consideration of the religious implications of man's exploration of the heavens, "The advent of the space age cannot one whit alter this crisis character of existence, but can only accentuate it and bring home to us in most striking fashion who and what we are and how much we need the God who made us."[2]

None of the issues that dominated theological reflections on the space age—the epistemological status of Christian cosmography, the implications of quantum advances in man's technological capacity for his relations with God, and the compatibility of religious doctrine with the discovery of intelligent extraterrestrial life—was original to the era. Into each of these debates, however, the space age introduced significant new elements: man was entering the heavens, not just studying them from the ground; he was now able to attain a synoptic view of Earth, analogous to the omniscience conventionally attributed only to God; and he had developed the means with which to initiate contact with other worlds, no longer confined to simply watching the skies and waiting for those worlds to come to him. As Heinecken observed, the novelties of the space age would serve to direct attention back to the basic question and then on to the answers of theology, for that was his reason for writing his book. But although theologians had achieved a symbiosis between faith and "objective uncertainty," would members of the laity want to do the same? Many, as it turned out, preferred to be assured that heaven could be situated in real space and real time, that man was still subordinate to the caring authority of God, and that Earth remained a site of special creation. The space age thus contributed to the perception among ordinary churchgoers that theologians, far from being the intellectual champions of faith, had already sold it down the river—in return for nothing more than a soulful sense of estrangement. This perception produced, not a recession of religious belief, but rather the further dissolution of theological authority over the field of religious thought. By the closing of the space age, popular cosmologies were very different in content and character—more esoteric and eschatological—from those anticipated by most theologians ten or fifteen years earlier.

Space-age theology was rarely systematic, in the formal disciplinary sense of the term. Aside from Heinecken's sustained reflections, there was no coherent whole constructed from its themes. There was not even a fully joined debate: many of those offering their thoughts seemed oblivious to what had been said on the very same subject by others before and around them. Only a handful of contributions extended over a few thousand words, and those that did—again, Heinecken excepted—were manifestoes rather than monographs, making the case that theology should engage with the advance into space rather than developing a detailed theological response of their own.[3] This hardly cancels the need for our attention, for it is possible to tender a broadly similar characterization of another, much more notorious strain of American theology that emerged in the 1960s: the theology that argued, on a variety of grounds, that God was dead.[4] Indeed, the two drew inspiration from the same sources and shared much the same fate, for by the end of the decade it was clear that in the view of most Americans God remained a certain and objective fact, well and truly alive, at home in his heaven and at work in the cosmos.

Christian Cosmography in the Space Age

For most theologians in the late fifties and early sixties, that man was propelling his machines and then himself out into space presented no immediate challenge to modern Christian cosmography. In comparison with other religions, Christianity had more readily permitted itself to imagine its God in material and spatial terms, his relations with the natural world incompletely constrained by the concepts of both transcendence and immanence. Christians could think of their God as existing beyond the natural world. Alternatively, they could think of him as existing within the natural world as a kind of divine essence, but as often as not he was seen as its overlord too, physically inhabiting its commanding heights and participating in its history. This was God as a Louis XIV of the skies, the gift of his grace expressed in the quality of proximity, bestowed and withheld through motions of ascent and descent.[5] For those who paid attention, the notion of a three-tiered universe, with heaven above and hell below, was reduced to the status of metaphor by the Copernican revolution: if the earth was not the center of the solar system, then there were no qualitative distinctions to be drawn between the different regions of space and no suitable physical location where God and his heaven could abide.[6] The point was further enforced by the progressive scientific revelations of an increasingly larger universe, containing many billions of stars, in which a God said to reside principally in a heaven up above the earth could be regarded as no more than a local deity, and by critical historical approaches to scripture, which, in attempting to identify the essence of the Christian message, consigned elements of its surrounding tradition, including the spatial reality of heaven, to the category of myth.[7]

By the beginning of the space age, theologians no longer conceived of a God who dwelled in the heavens and descended from there to intervene in human affairs. Moreover, Enlightenment science, with its insistence on providing naturalistic explanations for phenomena in the natural world, had so contracted the field for divine intervention that any God operating in the remaining gaps was in danger of appearing a little unworthy of the name. Hence, perhaps, the three primary orientations of religious thought in the early twentieth century. The first orientation was toward a natural theology that identified God as both immanent and creative in a process of "becoming" that linked all entities in the world.[8] The second, in dramatic and deliberate contrast, favored a neo-orthodox insistence on the extreme otherness of God and the impossibility of discovering him in the natural world.[9] Finally, there was an emphasis on the immanence of God, replacing the language of height with that of depth by asserting him to be "the ground of being" rather than an actual being in himself and defining his influence on the world in terms of the correlations to be drawn between the message of Christian scripture and the existential condition of man.[10] Among English-speaking theologians in the 1950s and early 1960s this third orientation—given systematic form by Paul Tillich and repackaged for broader consumption by John Robinson, bishop of Woolwich, in his book *Honest to God*—was probably the most influential.[11] Robinson made it clear that the existence of heaven was no longer a matter for debate. Although Christians might still speak of a God "up there," they did so in full consciousness of the figurative quality of their language. The work of theology had to be directed elsewhere, to the purpose of providing a meaningful alternative to residual popular notions of a God "out there," transcendent and supranatural; otherwise, religious faith might dissolve entirely in the face of modern science and its explanatory powers.[12]

Whether such theology was presented as a refinement or viewed as a retreat, its passage to popular acceptance was unlikely to be easy. All these recent variations—identifying God's work as "process," defining him as either "radical other" or "the ground of being"—functioned to withhold the traditional consolations of faith. Indeed, if God was not "with us," performing the role of protector and guide, then what was the point of paying him heed? One could plausibly begin a journey of spiritual inquiry adhering to any one of these positions, noted the death-of-God theologians, and ultimately finish it adhering to theirs.[13] It was this sort of intuition that lit the fires of controversy around Robinson's book, along with the fact that for many ordinary Christians it came out of the blue.[14] They had not been aware of how much theology had changed. As the sociologist Jeffrey Hadden later observed, "Wherever the theologians have gone during this century in search of God, most laity and a large proportion of the clergy have continued to live with the conception of a deity who is either 'up there' or 'out there.'"[15] Moreover, the two dispositions were probably synonymous. If there was still, as Robinson acknowledged, a widespread belief in a God "out there," then it followed that there must also be a conception of where "out there" he actually lived, and the tug of symbolic tradition, as well as the continuing need to express a sense of the elemental distance between God and man, left the ordinary Christian with his or her eyes lifted to the sky. At moments in his analysis Robinson seemed to admit the point: "This picture of a God 'out there' coming to earth like some visitor from outer space underlies every popular presentation of the Christian drama of salvation, whether from the pulpit or the presses."[16]

The space age disclosed these discrepancies between the cosmographies of the theologians and the lay believers. The rival Soviet space program had become explicitly identified with the Soviet regime's renewed campaign against religious institutions and belief.[17] That their spacecraft had encountered neither God himself nor any of his angels was asserted by Soviet scientists and cosmonauts to be evidence that such supernatural beings did not exist.[18] Nikita Khrushchev, the Soviet premier, declared mischievously that it had been the express mission of the cosmonauts Gagarin and Titov to investigate religious claims of a heavenly paradise, and they had found nothing.[19] Those in the West who were versed in modern theology, of course, were able to respond to these contentions with a measure of serenity. As C. S. Lewis observed, "If God—such a God as any adult religion believes in—exists, mere movement in space will never bring you any nearer to Him or any farther from Him than you are at this very moment."[20] Yet it was clear that not all believers were so well prepared. According to one commentator, writing in *Christian Century*, many pastors, not wanting to challenge the traditional piety of their parishioners, had failed to provide them with a concept of heaven that could survive the space age and stand up to Soviet mockery.[21] For evangelicals, who were more inclined to identify the precise wording of scripture with God's own truth, the problem was particularly poignant and called for careful exegesis. "We must be factual and historical in our proclamation of the events in which God was savingly revealed to men," cautioned one minister in *Christianity Today*, "but avoid suggesting that the divine world can itself be located in space and time. . . . The angelic worlds from which the Annunciation broke upon our earth must not be confused with some portion of discoverable space."[22] The loyalty of many laity to medieval mental maps of the Christian cosmos continued to perplex religious thinkers into the high summer of the space age. As one theologian reflected on the eve of the first moon landing, "A significant portion of the faithful clings to these old concepts: heaven as a 'place,' and God as a gray-bearded old man. 'Popular piety' of the church is certainly not present in the theology of any importance, but it darn sure is in the blood of the church." Still, he predicted, the mission would finally force a change in language: "Things theologians have been warning us about for a generation are going to become more apparent—the old 'up-and-down' of traditional religion, for instance."[23]

Yet, by the early 1970s it was apparent that whatever challenges had been presented to traditional cosmography by the exploration of space had not been universally resolved through an embrace of the concept of an immanent Christian God. Those for whom the space age had dissolved the sacred heavens could direct their devotions instead toward a sacred earth. In doing so, they might draw on the more digestible elements of process theology and revive, in the form of ecology, older Christian notions of a great interdependent chain of being, albeit without the same emphasis on a divinely appointed and hierarchical natural order. They were also profoundly influenced by animistic cosmologies encountered in the study of pre-Christian and Native American religious traditions. "At the macrocosmic level," declared Tim Zell, founder of the Church of All Worlds, one of the more prominent neopagan groupings, "we recognize that the entire Earth is a vast living entity: Mother Earth, Mother Nature, The Goddess." Divinity was also an attribute of the microcosmos. It was defined as "the highest level of aware consciousness accessible to each living Being, manifesting itself in the self-actualization of that Being. Thus, every man, woman, tree, cat, snake, flower, or grasshopper IS 'God.'"[24] At work in such beliefs was a conception of immanence, but it was hardly the same as what had informed the thought of Robinson and Tillich.[25]

Although reverent attitudes toward the earth were certainly not original to the long 1960s, studies both of the neopagan religions and of the secular environmental movement tend to identify the period as one that witnessed significant new concentrations of interest and activity.[26] That this new engagement with the earth, religious or otherwise, developed just as man was hoisting himself into space was largely a matter of poetic coincidence. It was not the space program that inspired Rachel Carson to write *Silent Spring*. When the adventures of NASA's astronauts did attract notice in the channels of environmental debate, the response was often critical: there was something decadent about departing on such jollies when the home planet was being poisoned, polluted, and turned into desert. Yet the perspectives provided by the exploration of space did help to compel a new appreciation of the earth and, for some thereafter, its sacralization too. What the manned and unmanned missions encountered in outer space—on the moon, Mars, and Venus—were worlds that were, to all measurable extents, dead. For many Americans, the romance of space travel quite probably began to wither the very moment that the *Apollo 8* astronaut William Anders, in the first-ever television broadcast from lunar orbit, compared the color of the moon to that of "dirty beach sand."[27] In Anders's famous photograph *Earthrise*, of a partly shadowed blue-and-white earth appearing over the desolate, grey lunar horizon, it is the earth, and not the moon, that appeals more to the eyes. Contrary to many expectations, the space age would yield no repetition of 1492; it would open up no lush, green paradise in which mankind could make a new start, construct a new home.[28] Neither the moon nor Mars seemed hospitable to human habitation, while Venus, named for the Roman goddess of love and beauty and once imagined to be an Edenic sister planet to Earth, turned out to have the sort of atmosphere and climate usually synonymous with hell.[29]

The Church of All Worlds was in this sense a space-age religion twice born. It originated in the early 1960s with a group of friends at Westminster College, in Fulton, Missouri, who were inspired by Robert Heinlein's science-fiction novel *Stranger in a Strange Land*. In the novel, a mission to Mars returns with Valentine Michael Smith, the son conceived by two Earth astronauts during an ill-fated expedition twenty-five years earlier. His parents dead, Smith had been raised by Martians. On Earth, he draws on the Martian religion, especially its belief in the oneness of all living things, expressed in the affirmation "Thou art God," to form his own church, the Church of All Worlds.[30] Over the course of the late sixties the nonfictional Church of All Worlds became more expressly ecological in its philosophy and placed more emphasis on the divinity of the earth itself than on the possibility of a cosmic religion.[31] In this move it reflected, albeit in magnified form, trends the space age was encouraging within the broader culture. Images of distant Earth taken during the Apollo missions to the moon, which, in the words of Archibald MacLeish, allowed mankind to see the planet "as it truly is, small and blue and beautiful in that eternal silence in which it floats," evoked both its fragility and a quality of presence, stimulating rhetorics of reverence as well as care.[32] On the first Earth Day, organized in April 1970 to demonstrate grass-roots concern about the state of the environment, the *NBC Nightly News* concluded its coverage with one of the Apollo photographs of Earth, shown in complete silence.[33]

Furthermore, space exploration—oriented as it had to be in its early encounters with other celestial bodies toward a panoramic comprehension of each—generated perspectives that were eventually redirected back to encompass the earth, to cast it as a unified life-world, even something that was effectively a living organism itself. It was an intuition shared by the Church of All Worlds and James Lovelock, a British scientist who had worked for NASA analyzing the atmosphere of Mars. They even gave it the same name, Gaia, after the Greek goddess of the earth.[34] But the thought was most cleanly expressed for a popular space-age audience by Anne Morrow Lindbergh, wife of Charles, writing in *Life* magazine after witnessing the flight of *Apollo 8* and reflecting on the new perspectives it had allowed: "For life, this rare and delicate essence, seems to be, as far as man's vision now extends, primarily the property of earth, and not simply life of man—life of animals, birds, butterflies, trees, flowers, crops. All life is linked. This is what makes up 'the good earth.'"[35]

It is easy to exaggerate the significance of alternative spiritualities in the religious life of Americans in the 1960s.[36] Probably the Church of All Worlds was not even a church in all of the US states; in 1973 its membership totaled only seven hundred.[37] The conception of the natural world and of the earth itself as manifestations of the sacred had wide cultural purchase; it was nurtured by academic inquiries into comparative religions and was encountered most commonly, as the seventies began, in ersatz pop renditions of Native American spirituality.[38] But it was not the turn to belief in the immanent divine, Christian or non-Christian, that was most notable in the religious culture of the late space age; it was the return to a belief in a transcendent God. If conjectures of immanence offered the consolation of authenticity, of a theology attuned to man's existential condition or to his place in a sacralized natural order, transcendence made a promise of justice and aid. In a time when the problems that man was encountering seemed to derive from his own constitution—from his propensity to selfishness, violence, and pride—authenticity was not much of a comfort. Hence the investment of hope in a power and an authority beyond.

The evidence of such a turn, or return, is dispersed but compelling. Intellectuals, here and there, started to remind themselves that while the natural sciences could yield satisfactory secular explanations for what happened in the natural world and the social sciences could find correlations between belief in the supernatural and the presence of certain social and psychological needs, the methods and instruments used in these disciplines were incapable of proving conclusively that the supernatural did not actually exist. Furthermore, these intellectuals noted, even in contemporary secular society it was possible to observe the continued appearance of what Peter Berger called "signals of transcendence," gestures that expressed a faith in an overarching cosmic order, in "a reality that is truly 'other.'"[39] Meanwhile, in the wider spiritual marketplace of the late 1960s the best business was being done by conservative and evangelical churches, that is, those who claimed fidelity to the Christian cosmological tradition and placed most emphasis on a personal encounter with God.[40] Opinion polls suggested an American religious imagination that was increasingly ripe and vivid, with broad affirmations of belief in the existence not just of a God of creation, salvation, and judgment but also of angels and devils, heaven and hell, and life after death.[41]

In the late space age, therefore, many Americans refurbished their cosmos with layers and contents that theologians had long considered antique. What was happening in outer space did not direct or determine this renewed apprehension of the supernatural, but it was not without effect either. In December 1969 an editorial in *Christianity Today* asked the question, "Has God Forsaken the World?" "Darkness has begun to descend upon the earth," it observed, "and the voice of God seems silent, his Spirit's restraining power lifted. Men grope in the darkness, reaching out to touch the hand that doesn't seem to be there, straining to hear the voice that doesn't seem to speak."[42] The following April another editorial, on humanism, complained about the "turning from God to man."[43] One month later, however, the channels of communication between the earth and the heavens were held to have been gloriously restored, a reversal of perspective occasioned by the safe return of the crew of *Apollo 13*, their craft barely functioning after one of its oxygen tanks had exploded on the way to the moon. *Christianity Today* commented that "millions of persons around the world were moved to pray for the imperiled astronauts, and God answered the intercession."[44] Billy Graham judged that the success of this call to prayer "might be used to bring a spiritual renewal that the world so desperately needs."[45]

What the space age also offered to those alert to the objective presence of the divine was the option of cognitive reinforcement through analogy. Just as a man traveling into space could be variously regarded as profaning the heavens or attaining a closeness to God, so too did an ambiguity persist over the religious significance of unidentified flying objects, or UFOs, reported sightings of which dramatically increased in number in 1973 and 1974.[46] For some evangelical writers, the UFO sightings were evidence of the work of demons and a portent of the coming tribulation.[47] Others were more sanguine, comparing contact with extraterrestrial beings to the gift of other embassies from God. For Billy Graham, the Christian cosmological tradition lent plausibility to conjectures of abundant life in the heavens: "It is hard to believe that we earthlings are alone in this spacious and wonderful Universe. Already we have received visits by creatures from outer space, including many angels and Jesus Christ."[48] And in turn, the revival of interest in UFOs became a sign of a "new openness to the supernatural," to the proposition that angels were still laboring on God's behalf and making their presence felt in the life of man.[49]

Finally, it was the space age that accelerated the assimilation within Christian apologetics of the "new physics," the body of hypotheses and concepts generated in the first half of the twentieth century, encompassing the theory of relativity and space-time curvature, quantum mechanics and the uncertainty principle, which transformed the ordered Newtonian universe into an altogether queerer place. By the time the space age began, much of what constituted the new physics was no longer really new, and some of its potential implications for Christian cosmology had already been identified. As E. L. Mascall observed in 1956, with the abandonment of the Newtonian conception of space as a uniform continuum came a renewed scope for theologians, if they so desired, to recant the existentialist doctrine that heaven could only be understood as a state of mind and return it to the skies, where it might conceivably reside unseen behind one of the bends in space.[50] But the movement of man and his instruments beyond the earth did create opportunities for theories to be further tested and refined and new advances to occur.[51] More importantly, perhaps, the Cold War competition in space, and especially the early Soviet successes, stimulated the expansion of high-school and college instruction programs in physics, the adoption of new curricula, and the revision of existing textbooks to reflect the current state of knowledge in the field. The result was that many more Americans in the late 1960s could claim an acquaintance with the strange geometries of the modern cosmos than had been the case a decade earlier.[52]

What the new physics appeared to license, in a way that Newtonian physics did not, was books like Barry Downing's *The Bible and Flying Saucers*, a sincere and, by the standards of ufology, relatively sophisticated attempt to subvert the normal patterns of exchange and use modern science to slow theology's retreat from the Christian cosmographical tradition. Downing argued that the elements of biblical narrative that described supernatural events no longer needed to be cast away like childish things or interpreted primarily as a correlate to the conditions of existence and being, for the space age had made them plausible again. If the many contemporary reports of UFOs turned out to be true, then perhaps they also offered an explanation for the parting of the Red Sea and the Ascension. If that was so, they were the equivalent of angels; and if they were angels, they must have come from heaven. It was here that Downing turned to the new physics, to notions of spatial curvatures and multiverses, to cautiously propose that "there may be something like a 'glass sheet' which separates our created universe from both 'heaven' and 'hell'" and that travel between the different realms would be possible by means of "a bend or warp in the space-time continuum, some kind of space tunnel."[53] He concluded: *"Heaven may be an entirely different universe right in the midst of us."*[54]

Of course, this was still very different from Dante's world. A wholesale reinstatement of the medieval cosmos exceeded even Barry Downing's aspirations. Yet his work exhibits quanta of both dexterity and ingenuity. It indicates how, when they made the effort to browse modern scientific thought and if they lifted carefully from its more enigmatic formulations and engaged imaginatively in their own labor of scriptural correlation, evangelicals could often find something they could use in their battle to stay the dissolution of the transcendental. Aspects of the same space age that had once been expected to confirm the necessity of the immanent turn by demonstrating empirically the archaism of traditional Christian cosmography were converted, as if through jujitsu, into a resource for evangelical hope. In a cosmos that remained hypercoded with material, if invisible, forms of the divine, heaven too could still be considered a place.

Space-Age Technology and the Authority of God

That the scientist and the seminarian have not always been at war is now something of a commonplace in the scholarly literature on the history of their relations. The Inquisition may have forced Galileo to abjure his own writings, and Wilberforce and Huxley may have shared some sharp exchanges, but the noise of such battles no longer attracts quite the same attention from historians that it once did. They have come to tell a more complex story—of a science nurtured in the early modern era by a belief in a God-given natural order; of a science inspired by assumed affinities between the study of nature and the act of worship; of Christian engagements with science characterized as often by adaptation and appropriation as by dismay and dissent; and, most simply, of religious scientists at work.[55] Across the scholarship that examines religion's encounter with technology, science's handy twin, the same interpretive frames tend to apply. There have been those who have discerned an ontological challenge to the transcendental in the utilitarian values that have shaped and speeded the development of technology, notably Jacques Ellul.[56] For others, however, Christianity itself, at least in its medieval and early-modern iterations, was the mother of invention. To labor may have been to pray, as Saint Benedict ruled, but man's best hope of recovering from the Fall and making himself anew in the image of God lay in laboring well, in experimenting with tools and machines, in creating by means of the "useful arts."[57]

In the United States in the mid-twentieth century, religious commentators on the subject of science and technology could find reasons for both optimism and apprehension. Living as they did in a nation that, *sui generis*, combined an advanced technological capacity with economic prosperity and an almost universal faith in God, it was easy to believe that a divine millennial design might be at work in the processes of research and development. After all, not every religious fundamentalist was a character birthed from the Appalachian gothic. A good many were employed as engineers in the emerging high-technology defense and aerospace industries of the American Sun Belt, and like the Benedictine monks, they believed that their professional endeavors were meet with the purposes of God.[58] Such confidence, however, was very highly conditioned. If a moral excellence abided in technology when it was directed toward the service of the divine, then moral horror attended the contemplation of what it might accomplish should it become detached from that service or, worse still, become enlisted in a campaign against it. This was no nebulous speculation. One had only to look east, to the Soviet Union, to find atheist engineers with an atomic bomb in their hands. Indeed, it was as much the need to secure the nation against this threat as any millennial hope invested in the actual achievements of its laboratories and workshops that persuaded evangelical Christians, who were traditionally suspicious of the state, to accept the expansion of government sponsorship of science and technology in the 1950s, and they were not alone in the fear that they were entering something of a Faustian bargain. Men and women of faith kept a watchful eye on the activities and aspirations of scientists and engineers throughout this period.⁵⁹ To the extent that theologians in midcentury America addressed themselves to the problem of sin, they usually did so with respect to the sin of pride, a theme with particular resonance in an era of "big science."⁶⁰ Likely, it would be his successful manipulations of nature that most tempted man into hubris, to aspire toward an image of God not as an expression of Christian piety but in order to effect an impersonation.

With the launch of *Sputnik 1* all these hopes and anxieties found a fresh and vital stimulus. In December 1957 the National Council of Churches identified the Soviet satellite, together with the intercontinental ballistic missile, or ICBM, as the herald of a new era in world history, "the nuclear-space age." The recent maturation of these technologies, the council noted, had engendered a sense of crisis: "Men and nations are reacting variously, in fear and hope, frustration and boasting, apathy and frenzy." It emphasized, however, that *Sputnik 1* at least was not necessarily a portent of doom: "We see possibilities for good in new dimensions of power, knowledge, and exploration of space, if used to enhance human life." But for that to happen, it was necessary to recognize that the present crisis was not merely military and scientific in nature: "It is moral and spiritual. It is related to faith and unfaith, the meaning of existence and history and the world, the understanding of God and his will, the nature of man and his destiny." Mankind was enjoined to remember that "God continues to rule over history with judgment and with grace" and that the new powers it possessed had been "discovered and developed under the Creator's sovereignty." Thus, if there was to be a resolution to the crisis, it had to begin with men turning to God "in faith, in prayer, and in action, that His will may be done in justice, righteousness, freedom, and peace—on earth and in the opening vistas of outer space."⁶¹

Throughout the late 1950s and 1960s, therefore, religious leaders and commentators found themselves equivocating between a celebration of space exploration as a fulfillment of the creative gifts that God had bestowed on mankind and a source of greater insight into the wider scheme of God's dominion and an apprehension that as man's capabilities increased to the point that he was able to contain the whole earth in his vision, he might forget the extent to which he owed his success to the sanction of the divine. Theologians may have been particularly sensitive to the scope for such a sin; after all, even some of their own had become sufficiently entranced by the promise of modernity to pronounce the death of God.⁶² Thus, as they praised the courage and technical skill of NASA's astronauts and engineers, those reflecting on the meaning of the space program often also invoked the Tower of Babel and the fate of Nebuchadnezzar.[63] Pope Paul VI applauded the first moon landing but cautioned that man was in danger of idolatrizing his own instruments, "perhaps to the point of madness."[64] Greeting the *Apollo 11* astronauts on their return to Earth, President Nixon declared that their journey had marked "the greatest week in the history of the world since the Creation," a statement for which he was gently admonished by Billy Graham, who suggested three "much greater" events: the birth, death, and resurrection of Christ.[65] Although the lunar mission's lack of formal liturgical content was regretfully noted by some religious observers, they were able to take at least modest satisfaction from the news that Buzz Aldrin had privately celebrated a communion after landing on the moon and from his reference to Psalm 8 during a live television broadcast: "When I consider thy heavens, the work of thy fingers, the moon and the stars, which thou hast ordained; what is man, that thou art mindful of him?"[66] As the theologian Paul Tillich had noted seven years earlier, the eighth psalm was a text that the space age had made newly ripe for exegesis. Though it asserted that man was "little less than God" and that he had "dominion over the works of thy hands," the psalm was addressed to a "Thou" from whom these faculties were derived, as a gift and a commission. In Tillich's words, the psalm affirmed that "we are brought into existence and formed by the same power that bears up the universe and the earth and everything upon it, a power in comparison with which we are infinitely small but our awareness of which makes us great among the creatures."[67]

For commentators identified with secular, progressive perspectives, the effort to envelop the space program within a devotional theme could itself be read as evidence of its transgressive ambition; after all, many an atrocious design had been dressed in the language of saints. They took as *their* text the personality of Wernher von Braun, then director of the Marshall Space Flight Center in Huntsville, Alabama, and, aside from the astronauts, the most prominent public figure at work in the program. Von Braun was strongly associated with the exultation of profane authority and the enabling of abomination as a result of his service in the Third Reich. Soon after his arrival in the United States in 1945, however, he became a born-again Christian, a conversion that most American evangelicals seemed content to take at face value, for it gave them the assurance that as the nation entered the space age at least one of its principal engineers was building his machines with God in his thoughts.[68] As von Braun informed the International Christian Leadership World Conference in July 1965, "The sooner we turn our eyes to Him for help, the better it will be, for ourselves, as well as for the objectives we are pursuing."[69]

The suspicion remained, however, that God was not just in von Braun's thoughts but also in his sights. The Italian journalist Oriana Fallaci brought to her interview with von Braun a memory of her family's wartime resistance to German occupation. He carried with him "a slight scent of lemon," a scent identical to that of the disinfectant soap used by German soldiers in Italy. To Fallaci, it was a visceral reminder of evil and terror: "Sharp, pungent, almost like a gas that penetrates through your nostrils right into your heart and brain." Yet she found herself beguiled: "For half an hour I made myself dislike him. To my utter astonishment I found myself feeling just the opposite." Von Braun, she wrote, possessed "a demoniacal ability to influence whoever is looking at him and listening to him."[70] For Norman Mailer too, von Braun evoked the diabolic: "Immediate reflection must tell you that a man who wishes to reach heavenly bodies is an agent of the Lord or Mephisto. In fact, Von Braun with his handsome spoiled face, massive chin, and long and highly articulated nose, had a fair resemblance to Goethe." Mailer noted a comment made by von Braun in a newspaper interview: "Through a closer look at creation, we ought to gain a better knowledge of the Creator." This was Mailer's exegesis: "Man was voyaging to the planets in order to look for God. Or was it to destroy Him?"[71]

Yet it quickly became clear that whatever their ambitions, von Braun and his associates had not conceived a new cosmology out of the first lunar landing; rather, they had conjured at best a magical spell that dissolved almost as soon as *Apollo 11* returned home. For Mailer, indeed, the magic had ended with the "machinelike" rhetoric with which Richard Nixon had addressed the astronauts on the moon.[72] "America," he observed, "was applauding Armstrong and Aldrin, and the world would cheer America for a day, but something was lacking, some joy, some outrageous sense of adventure. Strong men did not weep in the street nor ladies copulate with strangers." The Apollo program "had failed to become a church for the new age. She was only a chalice for the wounded bewildered heart of the Wasp, a code of honor for corporation executives hitherto bereft of pride." The "true product" of American technological civilization, Mailer asserted, was not von Braun's rocket with its claim on the heavens but the vending machines in the press enclosure at Cape Canaveral, "absurd machines poorly designed and abominably put together."[73] There was nothing in such "shoddy" technology with which man could speak to God.

Mailer was hardly alone in his observation that man's successes in space contrasted starkly with his record elsewhere. In the words of the *New York Times*, "It is by now almost a platitude to contrast the fantastic efficiency of the Apollo program with the ineffectual approaches the country has made to the poverty and malnutrition of its least fortunate citizens, to the alarming decay of its cities, to the sad decline of its public services, and to the pollution of its air and water."[74] There may have been some consolation in the thought that the space program could offer a model—combining technocratic management, technological innovation, and institutional idealism—for the pursuit of solutions to the chronic problems of the earth, but it was not much.[75] NASA itself could muster little enthusiasm for the task of generating terrestrial applications for its technologies.[76] Thus, the field was left to those for whom grand technological ventures like the Apollo program contributed to the problems of the modern world rather than to their solution. NASA's reach for the stars had, at the very least, diverted attention and talent from the challenges of alleviating poverty, cleansing the nation's waters, and defouling its air. But it was more broadly implicated in the crises of the era. "The same technological impulse that is carrying Apollo 11 outward to the moon," *Newsweek* noted, "is also threatening the home environment." The space agency, after all, was but the most glamorous incarnation of a transportation industry responsible for much of the effluence in the atmosphere.[77] Even on the morning of its greatest triumph, then, the promise of space technology was beginning to congeal; it had handed man the moon but perhaps also helped to cost him the earth.

It was not an irony that NASA would ever learn to savor: the fabulous quality of what it had accomplished on the moon was precisely what had dramatized the problem of priorities, drawn attention to its accounts, and made an issue of its means. In other ways too, equally unintentional, NASA would offer assistance to those who were anxious to convey what was at stake if mankind did not turn its eyes from the skies and look after its home planet instead. It was Buckminster Fuller who first invented the notion of a "spaceship Earth," on which life was sustained as a consequence of the synergy between its multiple interlocking parts but which, in order to slow the process of entropy, also required the careful conservation, recycling, and regeneration of resources.[78] The metaphor acquired more salience as the Gemini and Apollo programs matured, as missions were extended in duration and crews ventured further out into space; they traveled in craft that indeed had to serve as a surrogate Earth. "The sophisticated technology required to create the Apollo spaceship," *Newsweek* observed, "has done much to demonstrate the ecological principles of 'interrelatedness' in a closed system. NASA has spent billions to provide the moonbound astronauts with the ecology of the 'good earth'—pure air, pure water and careful disposal of waste. Each spaceship, in effect, is a model and a reminder of what earth should be like."[79] In due course, moreover, the jeremiadic mode of environmental critique would find in the space program, and specifically in *Apollo 13*, a compelling representation of what the earth was actually like, not least because the emergency occurred on the very eve of the first Earth Day. As Eric Sevareid commented on the *CBS Evening News*: "So the three astronauts head home across the desert of space, their oxygen and water running low. Perhaps the story will be seen one day as a parable. This earth is also a spinning spaceship, all of us are astronauts, and our oxygen and water are also diminishing. But we have no place to go."[80]

In the context of such apprehensions, the photographs of Earth taken during the Apollo missions became endowed with eschatological meaning. Small, alone, and watery blue, in these images Earth seemed more vulnerable, contingent, and provisional than it did when it occupied the horizon and supported man from beneath his feet. The same was true of man himself. As the astronaut Jim Lovell noted during the flight of *Apollo 8*, no trace of man's existence on Earth could be discerned from outer space.[81] Meanwhile, in the iconic photograph taken during the same mission, Earth rises above the surface of the moon but not, perhaps, by quite enough. Here is a lovely, living planet, but positioned below it, as though exerting a gravitational pull, is an ominous vision of what it might become: arid and desolate with no canopy of cloud and atmosphere to protect it from eternity.[82]

Beginning about 1970, there was a noticeable intensification of public interest in eschatological themes.[83] The bestselling nonfiction book of the seventies was Hal Lindsey's *The Late Great Planet Earth*, which surveyed contemporary world events—rising tension in the Middle East, the creation of the European Economic Community, and the increasingly frequent occurrence of natural catastrophes—and found in them evidence of the impending "end-times" as prophesized in the Bible.[84] Some editions of the book featured on their cover an image of a spherical Earth of the kind that had only become accessible in the course of the late space age, but here it was depicted in flames. The space age, of course, generated its own apocalyptic visions. Literary and cinematic fictions elaborated existential perils from the contingencies of space travel itself, especially the prospect that probes venturing to other planets might bring radioactive or biological contaminants back with them to Earth. Alternatively, they cast astronauts into the future as the chief witnesses to the world's dystopian fate.[85] UFOs, meanwhile, were open to a range of eschatological anticipations; they were read both as the work of demons—thus as a sign of the tribulation—and as the means by which the saved would ascend to heaven during the Rapture.[86] The knowledge that the sun would eventually die, and with it life on Earth, had long underpinned the conviction of rocket pioneers and advocates that a spacefaring capacity was essential; in its absence there was no hope of mankind's long-term survival as a species.[87] But, given the time frame involved—about five billion years—this was not an argument likely to win the case for expeditious development. In the early 1970s, however, it was dusted off and put to work in the context of concerns that through the progressive depletion of nonrenewable natural resources by an ever-enlarging population and the exorbitant ecological effects of industrial pollution, mankind itself was rapidly making the earth uninhabitable.[88] If the limits of sustainable growth had indeed been reached on Earth, asserted a few visionary thinkers, then Earth should be left behind: existing technologies were sufficient to permit the construction of large, rotating platforms in space and to establish on them pleasant, productive colonies powered by solar energy.[89]

What technology on Earth had wrought, technology in the heavens would fix. It was not to be, at least not in this particular space age. It was difficult to envisage the success of any humanly engineered equivalent of the Rapture when it took NASA thirteen years to move the space shuttle off the drawing board and into orbit. Man, and perhaps especially man in space, was no longer to be considered sufficient unto himself. Hence on 14 April 1970, during the travails of *Apollo 13*, the passage of a resolution by Congress calling on Americans collectively to pray, at 9:00 that evening, for the safe return of its astronauts.[90] "I think more people prayed last week than perhaps have prayed in many years in this country," observed President Nixon when the emergency was over.[91] "God has heard us," wrote 290 citizens of Grafton, North Dakota, in a telegram to the White House. "The events of these last few days have again taught us what wonderful abilities God has bestowed upon man and it has also taught us we are ever continually dependent upon him."⁹² The humanist confidence of the early sixties had buckled and left many Americans, when they thought about space, in a posture of genuflection. In December 1972, after its final lunar mission, Nixon offered the following meditation on the Apollo program: "Can we look at the record of 24 men sent to circle the Moon or to stand upon it, and 24 men returned to Earth alive and well, and not see God's hand in it? Perhaps, in spite of ourselves, we do still live in an age of miracles."⁹³

Yet, if the culture was humbled, it did not completely revert back to the piety of an earlier generation, as though the fall of the space age had simply canceled out all the effects of its rise. The environmental crisis had given a new critical twist to the insight that technology could encompass a Christian purpose. According to Lynn White, the remorseless instrumental exploitation of nature by man took its license from the distinction drawn between the two in the biblical account of the Creation and from the assertion in that account that man's dominion over nature had been granted by God.⁹⁴ By the turn of the seventies, however, Christian theologians had begun to respond to such critiques with the proposition that the problem was not so much the doctrine of creation as its interpretation and application. Man had forgotten that his dominion over the natural world was a gift held in trust; it was a stewardship, not an open invitation to plunder. If the conception of stewardship could be revived, then Christianity would bestow a vital, transcendental purpose on the task of saving the earth.⁹⁵ This, in turn, would allow certain forms of technology, including space technology, to be rehabilitated as the instruments of responsible dominion. Such a prospect was consistent with NASA's efforts in the early 1970s to redefine itself as an "environmental agency" and later to establish an important role for its satellites and other space platforms in the field of environmental science, culminating in its announcement of a "Mission to Planet Earth" in the late 1980s.⁹⁶

In addition, the same nation that had made a bestselling author of Hal Lindsey likewise proved remarkably responsive to the theories of Erich von Daniken, who proposed that in prehistoric times advanced extraterrestrial beings had landed on Earth, laid the foundations for human civilization, and implanted within man's consciousness a storehouse of knowledge that, its cues released in the manner of a genetic code, determined the rate and trajectory of his scientific and technological progress.⁹⁷ For von Daniken, the space age not only served to accustom human thought to the prospect of interstellar travel and encounters with intelligent alien life, thus justifying his contention that such encounters had occurred in the past; it was also the final manifestation of the extraterrestrial design. Man had been programmed to return to the starry skies from whence he had come. Like Barry Downing, von Daniken looked in the Bible and found descriptions of what he inferred to have been the visitations of spaceships, but in contrast to Downing, he did not regard such ships as devices of an ontological God. Prehistoric man had attributed divinity to those at the controls because there was simply no other way to express the radical distinction between the powers they possessed and his own. In the early years of the space age, Carl Jung had explained the proliferation of reports of flying saucers in terms of an unconscious desire to locate an alternative source of salvation and wholeness at a time when, given the authority of humanistic science, no one could expect an intervention from heaven.[98] What von Daniken offered at the end of the space age was a similar sort of comfort: God may not have been God, but that was okay, for man, perhaps, was still more than just man.

Christian Doctrine in the Space Age and the Plurality of Worlds

As they ran their calculations of the likelihood that life existed on other worlds, astronomers could reasonably affect an indifference to the innovations of the space age. After all, between the Copernican revolution and the mid-twentieth century, and without any aid from astronautic excursions, astronomical observation had displaced Earth from the center of its solar system, and the solar system from the centre of its galaxy, and revealed the dimensions of the universe to be so prodigious that simply looking at the night sky could become an experiment with vertigo. "What were we doing when we unchained this earth from its sun?" Nietzsche's madman asked in 1882. "Are we not continually falling. . . ? Aren't we straying as through an infinite nothing?"[99] For astronomers, the dizzying scale of the modern cosmos made it quick with biological possibility. By the mid-1950s the principal astronomical objection to the belief that in a universe containing many billions of stars there must be an abundance of life-worlds other than just that of Earth seemed to have been resolved. Confronted with the proposition that the sort of planetary system to which Earth belonged might actually be very rare, even unique, astrometric researchers had begun to produce evidence of the movement of planets around nearby stars.[100] Although final confirmation of the existence of such systems did not arrive until the 1990s, the evidence was sufficiently suggestive to enforce a consensus among astronomers that there were millions, perhaps billions, of locations in the universe on which life might have made a start, and this remained the consensus throughout the space age.[101] That NASA failed to detect biological activity on the two most proximate of these locations—Mars and Venus—certainly prompted some scientists to reflect both that a good proportion of planets might not be conducive to life and that life, especially intelligent life, would not necessarily emerge even on those that were. Nevertheless, the overall numbers involved were still so large that even the most conservative set of assumptions continued to yield a calculus that favored a rich plurality of worlds.[102]

Nor did it take the space age to inaugurate serious discussions of the theological implications of such a plurality. The proposition that life might exist beyond the confines of the earth preceded the birth of Christianity, retreated before the geocentrism of Aristotelian philosophy, and then reemerged in the wake of the Copernican revolution.[103] In the eighteenth and nineteenth centuries, debates about the existence, nature, and meaning of extraterrestrial life engaged the fulsome participation of astronomers and many other intellectuals.[104] And what lent a particular import to such debates, what lent them their vigor, was the inclination of contributors to argue either from the premises of religious doctrine or to a point of doctrinal consequence. Some, like Thomas Paine, took the assumption of plurality and made it central to their denunciation of the Christian system of faith.[105] For others, like the philosopher, scientist, and priest William Whewell, the terms of battle were reversed: the conjecture that intelligent life existed on other worlds was incompatible with, and withered before, what could confidently be asserted about the providence of God.[106] For others still, the Christian scriptures were much less directive on the question of plurality than they were, for example, on the immovability of the earth, and if theology could accommodate Copernicus, it could accommodate extraterrestrials too. Indeed, as William Derham pointed out in 1721, a plurality of worlds was exactly what might be expected of the Creator, given his infinite power: "As Myriads of systems are more for the *Glory* of GOD, and more demonstrate his *Attributes* than one; so it is no less probable than possible, there may be many besides this which we have the Privilege of living in."[107] The debate across these diverse convictions reached no resolution. Only a confirmed human encounter with extraterrestrial life could produce the sort of new information that might persuade participants to change their positions; in the absence of such an encounter few did so.[108] Meanwhile, most of the major Christian denominations offered no doctrinal determination on the matter. Though other worlds were incorporated into the cosmology of the Mormon Church, the Vatican elected to keep its own counsel.[109] As one Jesuit father observed in 1952, "The religious and moral interest of mankind has not yet required such a pronouncement."[110]

Into the space age itself, it was not difficult to find religious commentators who considered inquiries into the theological implications of a plurality of worlds to be a poor use of their time and intellectual energy. There were, after all, many issues of more immediate and practical significance for the future of faith and the church that merited discussion first. "We are trying to cross a bridge," declared C. S. Lewis, "not only before we come to it, but even before we know there is a river that needs bridging."[111] Still, even Lewis could not resist some further elaboration on the possible spiritual status of extraterrestrial beings and on the consequences of making contact with them.[112] Discernible in the corpus of astronautic and astronomical explorations that characterized the space age in science, and in the accompanying popular pageant of UFO sightings and imaginings, was a distinct quickening of man's physical and cognitive engagement with the cosmos. Contact, perhaps, was not very far away, whether it occurred as a result of man's own excursions to other planets, alien visitations to Earth, or, most likely, radio signals emitted into and received from outer space. "Intelligent life might be discovered at any time via the radio telescope research presently under way," noted one official report in 1960.[113] In the light of such possibilities, a number of commentators observed, Christians would be well advised to begin reflecting in earnest on the meaning of such an encounter for the doctrines of their church.[114]

To envisage a plurality of worlds was, in the idiom of astronomy, to make about the earth an "assumption of mediocrity." To encounter intelligent life on those worlds would probably be to make man mediocre too. In this prospect abided the substance of the space age's challenge to Christian scriptural tradition. Although the Bible did not expressly deny the existence of other worlds, it only told the story of God's creation of man on Earth. Hence the supposition, consecrated further by each century in which it remained unquestioned, that this was the only such story that could be told. The uniqueness of man was also inferred from the assertion in Genesis that he was made in God's own image. What, then, of man's status if intelligent beings were discovered elsewhere, for God, if he was an omnipotent God, must have created them too? What also of the quality of God's care for man on Earth in an infinite universe containing innumerable other sites on which he had created life? It was knowledge of the size of the cosmos that provoked Nietzsche's madman to cry "Where is God?" and to declare that he was dead.[115] In June 1959 the board of deacons of St. John's Lutheran Church in Stamford, Connecticut, used more moderate language to express a very similar apprehension in a petition submitted to their denomination's annual convention. Many Christians, the petition stated, "hesitate and falter" as they came to contemplate the universe and the notion of "thinking life on other planets"; they found it difficult to conceive of a God capable of maintaining a loving watch over man as well as the totality of the rest of his creation.[116] "If there are any gods whose chief concern is man," declared Arthur C. Clarke, "they cannot be very important gods."[117]

It was a sign of the prevailing hold of the presumption in favor of plurality that very few space-age theologians responded to such challenges with the answer that had been offered a century earlier by William Whewell: that the logic of Christian providence excluded the possibility of intelligent life on other worlds.[118] Instead, they took their stand with William Derham, asserting that the discovery of multiple sites of creation would simply provide further evidence of the power and glory of God and confirm the necessity of worship and praise. As the Jesuit astronomer Francis J. Heyden commented, "The more we know of the universe, the more we know of the works He has given and done. This brings us closer to God and closer to that perfect fulfilment: to know God." The existence of intelligent extraterrestrial beings, then, "would be a reaffirmation rather than a denial of God."[119] Moreover, a deity capable of creating an infinite universe was more than equal to the task of attending conscientiously to each of its parts. Indeed, as the Jewish philosopher Norman Lamm pointed out, for as long as a concept of divine immanence—of a Creator who still abided within his creation—was allowed to coexist with an emphasis on God's transcendent power, man on Earth could be reassured that he retained a place in the providential scheme. "Part of God's endless praise is that despite His loftiness and our lowliness, He is still concerned with every one of us—and every other rational sentient race anywhere."[120]

Of course, the argument that in a vast universe God could still care for man on Earth did not quite satisfy the conventional Christian expectation that man on Earth would be first in his thoughts. Theologians could try to present the proposition that intelligent beings on other planets might also have been made in God's image as a stimulus to wonder and praise, but there remained a sense that something fundamental was being surrendered. In particular, as Paul Tillich observed, there was the disturbing question of what a plurality of such worlds would mean for the status of the Incarnation, long identified as the center of the "Christian view of human history" and as the principal manifestation of God's regard for man.[121] Had he not, after all, so loved the world that he gave his only begotten son?

To these questions there were three possible responses, none of which entirely reconciled the scientific cosmologies of the space age with the avowal of man's justification in Christ.[122] Firstly, it was conceivable that there existed on other worlds intelligent beings who actually had no need of redemption, who lived in a state of grace, as Adam and Eve had before the Fall. If this were the case, however, it would provide a sharp corrective to the meaning usually ascribed to the Incarnation on Earth: Christ died not because man was particularly loved but because among the inhabitants on the various sites of creation he was particularly depraved.[123] Secondly, in the event that other planets hosted beings who had fallen in much the same manner as man, there was the prospect that the Incarnation of Christ on Earth was sufficient to have redeemed them too. This at least returned both man and Christ to the center of cosmic history, but it also raised significant difficulties. It took a very optimistic view of the technical scope for interstellar communication to imagine that the good news might work its effects outside the immediate neighborhood of Earth.[124] As one official report noted in 1968 to illustrate the improbability that the same journeys could be made by UFOs, knowledge of the Incarnation "could not yet have reached as much as a tenth of the distance from the Earth to the center of our galaxy."[125] Moreover, even if other worlds were to receive such news, this would not necessarily satisfy the terms of the Atonement. It mattered that the Son of God shared the same flesh as those he had come to redeem. His incorporation into man, which allowed his death, was precisely the source of man's salvation.[126]

Thus, for most theologians who considered the issue, a separate process of redemption—and, most likely, a separate Incarnation—would have to take place in each of the worlds where such redemption was needed.[127] This was not, they asserted, the heresy that it first seemed. As the Anglican theologian W. Norman Pittenger observed, it was often forgotten that for "the fathers of Christian theology" the life of Christ did not "exhaust the possibility of knowledge of God and relationship with him."[128] According to the Nicene

Creed, Christ was indeed the Incarnation of the Word, but it did not declare him to be the only possible one. After all, it was the essence of the Holy Trinity that a plurality of persons—Father, Son, and Holy Spirit—could subsist within the unity of God.[129] Hence, the Word itself—the Son of God—could be incarnate in forms other than Christ on many different worlds.

It was not in the style of the debate on the implications of the infinite spaces of the universe for Christian doctrine for participants to tender their thoughts dogmatically. Rather, they presented their contributions as useful in the sense that if it chose to attend to them seriously, the church might not be entirely unprepared for the scenario of extraterrestrial contact, but also as necessarily inconclusive: until such contact occurred, answers—like the cosmos—would remain open ended. To discuss these questions, wrote E. L. Mascall, was to demonstrate "how wide is the liberty that Christian orthodoxy leaves to intellectual speculation, and how many are the avenues that it opens up. Theological principles tend to become torpid for lack of exercise, and there is much to be said for giving them now and then a scamper in a field where the paths are few and the boundaries undefined; they do their day-to-day work all the better for an occasional outing in the country. Outings, however, are outings and work is work, and it is very important not to confuse them with each other."[130] In due course, indeed, there emerged an additional reason for discounting the urgency of these questions. Within a decade or so of the end of the space age, the assumption of mediocrity was exposed to an apparently substantive challenge on the basis of the "anthropic principle": that the emergence of intelligent life on Earth could not have occurred if the physical and biological conditions prevailing there had been more than marginally different.[131] Even in a universe containing many billions of stars, such conditions might well have been unique, and so too perhaps was man.[132] Inevitably, the anthropic principle was turned into an argument for design: over the vast reaches of cosmological time and space, what God had intended all along was the emergence of man on Earth.[133]

The anthropic principle did not supplant the assumption of mediocrity as the new point of consensus in cosmological science. But its rise to prominence was evidence of both the failure of the space age to fulfill expectations of an encounter with other worlds and the inability of theologians to provide Christians with a reason for keeping such an encounter at the center of their hopes. What the space age had actually revealed, wrote Rodney W. Johnson, an evangelical scientist employed on the shuttle program, was "the truth of the biblical account of creation" and "the uniqueness of man and our divine origin." This, he suggested, had been God's intention for man in space all along.[134] The assertion that there might exist millions of different incarnations of the Word across the universe now sounded a strikingly dissonant note within a national religious culture that by the late 1960s appeared to be focusing with increasing intensity on the fate of this particular earth and the person and promise of its own particular redeemer, Jesus Christ. The emergence of the Jesus movement within the counterculture, the growth of the conservative Campus Crusade for Christ, and the success of the album, theater, and cinema productions of the musical *Jesus Christ Superstar* were all indicative of the trend.[135]

By the mid-1970s, indeed, the contents of the space-age debate about the theological implications of a plurality of worlds had largely disappeared from memory. In 1978 *Time* magazine could assert that "major religious thinkers have yet to give serious attention to the issues posed by what some call 'exotheology' (the theology of outer space)."[136] An article in *Christian Century*, meanwhile, endeavored to "initiate some dialogue on the matter."[137] Periodically in the years since, similar calls for systematic study and discussion have been issued, but responses have tended to be sluggish and desultory.[138] Thus, in the long twentieth century it was the space age that prompted the most extensive, though still hardly systematic, theological engagement with the challenges presented to church doctrine by the prospect of an encounter with intelligent extraterrestrial life. As with the other principal themes of space-age theology, however, the answers theologians reached had less influence on the immediate currents of popular religious belief than the gravitational pull of an older, more familiar cosmos, ordered by a God residing in his heaven, maintained by his grace, and redeemed by his Son in the person of Jesus Christ.

"The horror of the Twentieth Century," Norman Mailer declared in his account of the first moon landing, "was the size of each new event, and the paucity of its reverberation."[139] Certainly, those who had predicted that the space age would stimulate dramatic transformations in Christian habits of thought—forcing heaven to become a metaphor, celebrating the maturing of man's ability to assume his divine inheritance by exercising dominion over the earth, and rendering provisional any doctrinal assumptions concerning the status of the earth within the scheme of creation and the singularity of the Incarnation of Christ—found at the end of the age that their prophecy had failed. Astrotheology remained a speculative prospectus, with more proponents than exponents: few were confident enough of a worthwhile return to devote their own intellectual careers to the task. Yet, though the space age may have disappointed theologians, it was not without consequence across the field of religious thought. It had encouraged some Americans to sacralize the earth. To others, it had affirmed that man, vulnerable both in space and on his own stricken planet, remained in urgent need of the providence of God. And for as long as the space age lasted, there continued to be anticipation of cosmological effects generated by the experience of spaceflight itself. By the early 1970s, indeed, the most significant attempts to define the religious meaning of the space age were being made, not by theologians, but by those who had actually voyaged into the realm beyond Earth: the nation's astronauts.

CHAPTER THREE

Into the Other World

Anticipations of Spaceflight as Religious Experience

"*What would you like to say first?*" It was late in the afternoon of 20 February 1962 when John Glenn found himself a quiet spot on the deck of the USS *Noa*, a navy destroyer. Holding a tape recorder in his hand, he started to answer a NASA questionnaire. An hour or so before, the crew of the *Noa* had hoisted Glenn's capsule, *Friendship 7*, out of the Atlantic Ocean, following its splashdown about 170 miles east of Grand Turk Island. *Friendship 7*, in America's first manned orbital spaceflight, had just carried Glenn around the world three times. Now, to the west of the *Noa*, the sun was beginning to set. Glenn had always savored sunsets—"God's masterpieces," he called them. Before his mission, the prospect of experiencing a new sunset with each of his orbits had stirred his imagination, and the spectacle did not disappoint. The sun—"as white as a brilliant arc light"—had shone through the prism of Earth's atmosphere to produce a spectrum of colors more vivid and varied than that normally witnessed on the ground: "greens, blues, indigos, and violets" as well as "reds, oranges and yellows." Then it had melted swiftly away "into a long thin line of rainbow-brilliant radiance along the curve of the horizon." And so, the mission over, what did John Glenn want to say first? "What *can* you say about a day in which you get to see four sunsets?"[1]

Glenn, then, only had the briefest of opportunities to reflect on what he had experienced aboard *Friendship 7* before his thoughts were wrested from him and inserted into the official record. NASA wanted its information as fresh as it could get it. Hence the questionnaire, and hence also Glenn's transfer that same evening via helicopter to an aircraft carrier and onward by plane to Grand Turk Island for two days of debriefing. Waiting impatiently behind the NASA technocrats stood the rest of America. Glenn had already enjoyed a measure of national celebrity before he joined the space program, having broken the record time for a cross-country flight in July 1957.[2] He was the most articulate of the original Mercury Seven and the one who was most attuned to the moral and cultural content of the public expectations that surrounded the astronauts. He was also the one who seemed most naturally able to reconcile the technical and existential demands of space-age pioneering with small-town American values, inherited from an upbringing that, as Glenn himself observed, could have provided the model for Norman Rockwell's entire *oeuvre*.[3] In addition, the launch of *Friendship 7* had been postponed many times over a period of two months, allowing anticipation of the mission to build and channeling the nation's attention toward its calm, resilient pilot.[4] And Glenn, once finally launched in his capsule, would confront a conspicuous hazard not encountered in the suborbital missions that preceded his own: a fiery reentry through Earth's atmosphere. During the flight itself, telemetry data indicating that the heat shield designed to protect him had come loose—data shared with the millions following the mission live on television—made that passage back to Earth seem even more fateful.[5] But Glenn survived, and as he talked to the debriefers on Grand Turk, national leaders worked to shape welcome ceremonials fit for the homecoming of this most modern of American heroes. His debriefing duties completed, Glenn returned to Florida. The drive from Patrick Air Force Base to Cape Canaveral became a twenty-mile-long parade through cheering crowds; at the Cape, he was greeted by President Kennedy, and together they went on a tour of the NASA launch facilities. After the weekend, Glenn traveled with Kennedy aboard Air Force One back to Washington, where he addressed a joint session of Congress. Two days later, he testified before the Senate Committee on Aeronautical and Space Sciences. The next morning, New Yorkers tossed thirty-five hundred tons of confetti and ticker tape onto Glenn's motorcade as it drove down Broadway.[6]

Some of the celebrations turned distinctly reverential. The *New York Times* commented that Glenn's appearance before the Senate committee "took on the aura of a revival meeting in a tent in Oklahoma."[7] The success of *Friendship 7*, the senators confidently asserted, had effected a "spiritual uplift" in the country at large, and they seemed to mean more than just an improvement in national morale. "I personally feel," Senator John Stennis told Glenn, "that you and your co-workers have been a great medium for revival of spiritual feelings and spiritual values." The senators were keen to know, indeed, whether the flight had been a spiritual experience for Glenn himself, for the astronaut had always spoken openly and without embarrassment about his faith. "I was wondering," said Senator Alexander Wiley, "whether you felt that God was up there as well as down here, and that in Him you lived and moved and had your being?"[8]

"Absolutely," Glenn replied, but his testimony before the committee was not entirely faithful to the analogy with revival or to Wiley's conjecture that his flight had involved a dynamic perception of God's presence and love. Each of the Mercury astronauts had passed through an intensive selection process designed at least in part to exclude anybody with a psychological disposition that might lead them to become overly excited by the experience of spaceflight. Glenn's religious convictions were wholly consistent with the premium NASA placed on stability. Contrary to the opportunistic assertions of some American evangelicals, he was not a "born-again" Christian.[9] He continued to hold to the mainline Presbyterian faith in which he had been raised, and his encounters with the cosmos during the flight of *Friendship 7* simply reaffirmed that faith rather than transforming it. They were encounters, moreover, that were both broadly predictable and distinctive only in quality or number, not really in kind, from those available on the ground. The three sunsets that Glenn witnessed from orbit were an obvious case in point. Due to the density of his capsule window, Glenn's view of the stars and planets had been no better in space than it would have been from a desert on Earth on a clear night.[10] Although he believed that the "hugeness" and "orderliness" of the heavens confirmed the existence of "a God, some Power that puts all this into orbit and keeps it there," his knowledge of those attributes of size and system remained essentially secondhand.[11] What Glenn maintained was a steady-state faith for a steady-state universe, and it was unlikely to be much affected by his excursion one hundred miles above the earth. So he informed the senators that he had not lapsed into prayer in the course of the mission, and he refused to claim that his experience had afforded him any special insight into the nature of God. If God had been with him in orbit, it was not because Glenn had journeyed any closer to his kingdom but simply because God was present "wherever we go."[12]

Looking back now, the particular combination of the communal rites of celebration that followed John Glenn's return and the rigor and reticence with which Glenn himself discussed the religious meaning of his mission qualifies the flight of *Friendship 7* as the last hegemonic moment of mainline American Protestantism. Ten years later a radical, expressive pluralism was at work across the field of religious belief and practice in the United States. By the early 1970s Americans appeared to be in the midst of another "great awakening," busily discounting the doctrinal and ritual claims of established religious institutions and asserting instead the primary authority of their own spiritual experience.[13] And just as the early manned space program had presented in the person of John Glenn the exemplary model of mid-century liberal Protestant man, its later iterations produced avatars compatible with both the variety and the seeking impulse of the new religious era. A more diverse cohort of astronauts, journeying into space in service of the lunar mandate established by President Kennedy and exposed in the course of those missions to a progressive series of almost entirely original experiences and perspectives—including distant views of the whole earth, vistas that declared the infinite depths of the universe, and traverses across the haunting landscape of the moon—returned home and out into a spiritual marketplace that seemed eager to discover whether they had been changed by their adventures. As the Apollo program drew to a close and discussions of its legacies began, news reports frequently invoked the examples of Edgar Mitchell and James Irwin in support of the proposition that spaceflight had yielded a spiritual return in the effects it had on the astronauts themselves and also in the messages that they in turn sought to communicate to others. "The Apollo veterans have become poets, seers, preachers," *Time* magazine asserted, "all of them evangelists for the privileged vision from space that Edgar Mitchell calls 'instant global consciousness.'"[14]

It is easy, however, to exaggerate the extent to which Americans were shifting their religious allegiances and seeking out new sources of spiritual insight at the end of the 1960s. Similarly, it is tempting to attribute to the experience of spaceflight an elemental power to transform religious consciousness that it does not possess. As C. S. Lewis had observed early in the space age, whether or not an astronaut found God in space depended very much "on the seeing eye"—who that astronaut was and what he already believed.[15] It also mattered what was actually seen. Not every hour on every mission offered an equal ration of wonder, the same promise of epiphany. Spiritual experiences in space, like spiritual experiences anywhere, were a complex compound of historical contingency and a sense of the unconditioned, of something that had traveled direct from the very ground of universal truth. A ticket into orbit, or onward to the moon, was not enough by itself to ensure that its bearer would come back a different man. Some Apollo astronauts reported that in matters of mind and spirit they had been changed by their missions hardly at all.

Contrary to *Time*'s assertion, then, not every veteran of the program felt a compulsion to evangelize. And even those who did found it difficult to maintain momentum beyond the duration of the space age itself or to construct a spiritual message that matched the ecumenical appeal of the secular adventure to the moon. James Irwin enjoyed a season of successful religious ministry, while Edgar Mitchell attained the status of a guru in the more cerebral precincts of New Age spirituality, but by the late 1970s neither could claim an audience commensurate with his original ambitions. They had recently assayed the heavens, but they could not evade the humbling rigors of the spiritual marketplace on Earth. Like many other prophets before and around them, they were left to narrowcast a vision conceived in universal terms. And their fate seemed to point to the withering of some larger social promise once expressed in the encomiums with which a committee of American senators had welcomed John Glenn in 1962: that the exploration of space would renew the spiritual life not only of the astronauts commissioned to undertake it but of the nation and the planet that had sent them on their way.

Conceptions of Religious Experience

There have been three major distinctive forms of religiosity in America: the evangelical, the "mainstream-denominational," and the metaphysical.[16] Each of these traditions retains its own characteristic model of religious experience: the sudden and intimate awareness of being charged with God's grace, in the case of evangelicals; the reverence and the sense of sanctity that accompany participation in the customary liturgical practices of a long-established church; and the intuition of a cosmic unity, of an integration of the individual mind with the energies at play in the universe at large, claimed by most American metaphysicians. For adherents of the traditions, these models of experience have been more than just another marker of religious identity. They express the principal tenets of belief—about the relation of God and man, God and the church, man and the universe—and they have also evolved into powerful behavioral cues. Indeed, it is most often by means of a religious experience that someone new to a tradition establishes his or her claim to belong.

From the late eighteenth century, the question of the relationship between the experiential elements of a religion and the integrity and authority of its doctrines and rituals started to achieve clear definition and to attract sustained attention from writers and thinkers in America and elsewhere. In part, this was a response to the vitality of the new "religions of the heart"—Pietism and Methodism—which challenged individuals to seek their own direct revelation of God and not to rely on the pharisaic structures and practices of the church.[17] But it also reflected a philosophical turn, directed in particular by David Hume and Immanuel Kant, that dismissed the possibility of a natural theology, of making rational empirical arguments for the existence of God. In combination with a rapidly evolving body of scholarship that historicized both the composition of the scriptures and the emergence of orthodox church doctrine, identifying them as the products of human rather than divine agency and inspiration, this critique of natural theology presented believers with a poignant problem. If they could no longer argue from design or trust in the infallibility of church and scripture, from what source was their faith to receive validation?

The answer lay in experience. Although Kant had declared that God could not be recognized in the world, his philosophy allowed for an intuitive grasp of a transcendent principle in matters of aesthetics and morality: there are certain objects that all men would view with pleasure, and our apprehension of a particular action as good or evil is drawn from an innate awareness of a universal moral law. Here was an opening for Christian apologetics. Why, if we can intuit beauty in an object, or evil in a deed, can we not experience a state of affection and know it as the work of God? For the German philosopher Friedrich Schleiermacher, raised in the Pietist tradition, "consciousness of being absolutely dependent" was the original and necessary source of religious knowledge, from which all else flowed—faith, action, and, as this consciousness formed the principal foundation for religious fellowship, the institutions of the church itself.[18] It was not just that religious experience could have a cognitive value, constituting a sign of God's grace; it was also autonomous and primary, the very ground on which the structures of religious doctrine and ritual and communal life were built. In the early twentieth century, when functional and reductionist explanations of religion arising out of sociological and psychological theory seemed to command broad intellectual assent, it was again to experience that theologians turned to refute the claim that religious faith was an epiphenomenon, the issue of other, more elemental social forms. For Rudolf Otto, in his seminal treatise *The Idea of the Holy*, the experience of the "numen"—of something so powerful and mysterious that one felt reduced in its presence to the status of a creature "abased and overwhelmed by its own nothingness"—was "the real inner-most core" of all religion and without analogues outside the religious life. The "numinous" state of mind, he wrote, "is perfectly *sui generis* and irreducible to any other."[19]

In America, the conviction that affective experience offered a distinctive and constitutive basis for religious belief drew support from a philosophical tradition that ran from Jonathan Edwards to William James, via the transcendentalists. Though the tradition originated with Edwards, it eventually left the sectarian fires of the Great Awakening far behind, progressively accommodating modes of religious experience that had developed well outside the cultural milieu of Protestant Christianity. The analytical challenge of abstracting the common essence of these experiences also encouraged an elastic conception of what might constitute the divine. Richard Bucke, writing at the start of the twentieth century, identified in sources as diverse as the Bible, Eastern mystical texts, the poetry of Walt Whitman, and the Koran a class of psychical events that he considered evidence of an evolving "cosmic consciousness." Bucke described his own sudden flaming realization "that the Cosmos is not dead matter but a living Presence, that the soul of man is immortal, that the universe is so built and ordered that without any peradventure all things work together for the good of each and all, that the foundation principle of the world is what we call love, and that the happiness of every one is in the long run absolutely certain."[20] Not long afterward, drawing on a similar range of sources, William James published *The Varieties of Religious Experience*, in which he argued that the origins of such experiences were a less productive subject for empirical study than their effects. Like Schleiermacher and Otto, James believed that religious experience created the basis for a religious life, the exemplary instance being the spiritual "genius" whose personal communion with the divine inspired the establishment of a new church or sect.[21]

Still, into the middle of the twentieth century reductionist assumptions continued to inform most scholarly discussions of religious experience, and they also worked to deter much substantial empirical research on the subject.²² In an era when the canonical texts of social science all asserted the fact of secularization, and when many liberal Protestant churches had traded in the goal of conversion for programs of social amelioration and civic reform, divining the nature of religious experience seemed an inessential intellectual task.

By the late 1950s, however, the question why, and in what form, people came to have an encounter with the sacred had assumed a new significance. For more than a decade evangelical preachers had been holding revival crusades in some of the nation's largest cities, with the attendant spectacle of audience members in their hundreds stepping forward, many of them in tears, to make a public commitment to Christ.²³ In 1962, 20 percent of Gallup poll respondents attested that they had experienced "a moment of sudden religious insight or awakening."²⁴ In a survey of members of Protestant churches conducted in 1963, 45 percent declared with certainty that at least once in their life they had been "in the presence of God," while 37 percent were sure that they had been "saved in Christ." "For all that religious experiences may be strange," the authors of the survey observed, "they are not unusual."²⁵ There was a growing intellectual challenge, moreover, to the standard sociological view that the perception of the sacred presence that defined such experiences was simply a projection of something else. In the theology of Paul Tillich, with its emphasis on the potential for all men to receive the "Spiritual Presence" and to become open to "the transcendent unity of unambiguous life," in Carl Jung's call for exploration of the "primal waters" of the "collective unconscious," and in Mircea Eliade's account of the indefinite repetitions of archetypes of the divine that were common to archaic societies, there was a shared conviction that an elemental and universal constituent of human culture was manifest in religious experience.²⁶ Meanwhile, wanting their discipline to contribute more to the world than just methods of behavioral adjustment or Freudian models of neurosis, humanist psychologists emphasized the development of human potential, or "self-actualization," drawing their data from empirical studies of individuals who enjoyed excellent mental health rather than from instances of psychopathology. Their research suggested that the condition of psychological well-being was frequently attended by "peak-experiences," moments of acute concentration and comprehension charged with a strong sense of belonging and love. The peak experience, they proposed, was the original, basic form of the religious revelation.²⁷

Another decade passed, and from it Americans seemed to emerge either in hungry pursuit of religious experience or aflame with religious experience recently procured. By 1979 more than a third of adults polled were attesting to "a life-changing religious experience."[28] In most cases this involved a conversion in which they had asked Christ to be their personal savior. There was also a turn toward experiential novelty, accessible through participation in the "new" religions of the East, native spirituality, nature worship, and the occult. What linked these spiritual movements was a belief in the fundamental unity of the cosmos: the smallest element and the largest element were implicated in each other; matter and mind could be involved in a mutual shaping; man belonged to the universe, not the universe to man.[29] The experiences they sought to kindle in their adherents shared a common essence, a consciousness of the whole, of the divinity of all. The religious counterculture was more conspicuous than ubiquitous. Even in the Bay Area of San Francisco, the principal social laboratory of the New Age, not quite 8 percent of residents had participated in yoga groups, while just over 5 percent had tried transcendental meditation.[30] But it is clear that over the course of the space age many Americans had drawn the promise of religious experience closer to their lives and that their quest for fulfillment of that promise had started to carry them to ritual spaces well beyond the enclaves of the mainline Protestant churches.

Exploration, the Environmental Sublime, and Divinity in Nature

A few minutes after John Glenn and *Friendship 7* had arrived safely aboard the USS *Noa*, President Kennedy walked out onto the White House lawn to speak to the press. "We have a long way to go in the space race," he observed. "We started late. But this is the new ocean, and I believe that the United States must sail on it and be in a position second to none."[31] As the phrase "new ocean" indicates, in common with NASA's own recurring invocations of America's frontier past, the proponents of the manned space program frequently cast it as a continuation of a long national tradition of discovery and exploration.[32] Certainly, what was anticipated in these journeys—by the astronauts themselves and throughout the broader public sphere—was consistent with the promise that had accompanied many earlier pioneers: the experience of an original relation with a strange, new environment. And that same tradition also offered a schooling in the possible effects of such an experience, from near-total dissolutions of self, to complex combinations of elation and horror, to efforts at integrating the old world and the world now encountered within a single cosmic scheme.

The earliest European narratives of the discovery of America are haunted by two questions: What is this place? Who am I? In the scale of its natural geography, the variety of its native peoples, and the abundance of its resources, America exceeded the European order of knowledge, and so the first response of the traveler was simply to stop and stare. "At the heart of the discovery narrative," Wayne Franklin has asserted, "stands the ravished observer, fixed in awe, scanning the New World scene, noting its colors and shapes, recording its plenitude and its sensual richness."[33] In such a gaze there was the quality of a communion, for what else could one be looking at besides the providence of God? But it also confronted the traveler with the contingency of nearly everything he had known before and by which he had measured himself. Although a prophetic motivation may have been at work throughout Columbus's career, most of the evidence suggests a turn toward an intense and apocalyptic mysticism in the course of his third voyage, when he coasted past the verdant shoreline of the Paria Peninsula, in the northeastern part of Venezuela, and surmised that he had located the Garden of Eden.[34]

Visions of a virgin paradise, ripe for the claiming, could also be confounded, and the despair that resulted would sometimes find expression in a rather different kind of epiphany. Stranded on Jamaica during a calamitous fourth voyage, Columbus wrestled with the apprehension that he had finally transported himself beyond the horizon of God's concern: "Solitary in my trouble, sick, and in daily expectation of death, surrounded by a million of hostile savages full of cruelty, and thus separated from the blessed sacraments of our holy Church, how will my soul be forgotten if it be separated from the body in this foreign land?"[35] A similar apprehension seems to have tormented Meriwether Lewis during his journey with William Clark across the Louisiana territory, as a landscape consistent with republican values of harmony and order—with the image of the garden—devolved into a wilderness, a ravaged and malign terrain of forbidding mountains and jagged cliffs, suffocating heat and violent storms, menacing beasts and relentless biting insects. Lewis descended into a depression, broke off his journal, and, soon after his return from the expedition, died a broken man, likely by his own hand.[36]

By the early nineteenth century the apprehension of a hostile environment often could be eased into an appreciation of natural grandeur by means of the language of the sublime. Licensing the sublime were emergent theories of an enlarged universe—Kant's "starry heavens above"—and a growing recognition that the vast prodigies of the earth's surface—mountains, volcanoes, and oceans—had been created by interminable natural processes, not by a vengeful God in the course of the Great Flood.[37] To gaze on these prodigies, therefore, was no longer to be reminded of man's sin and its consequence. Instead it was to receive an intimation of infinity. This provoked its own variety of dread, for in the depths of the night sky and in the power of a volcanic eruption lay certain evidence of the viewer's poverty and insignificance. But, asserted Kant, the work of the mind in recognizing that which exceeded its faculty of sense yielded simultaneously a "supersensible" pleasure: the mind that imagined what could not be measured delighted in itself.[38] Over time in the United States even this concept of the sublime, with its complex dualism of human capacity and incapacity, withered before the bloom of national self-assurance. Increasingly, in literature, paintings, and photographs, sublimity was evoked and then contained: the wilderness was ordered into landscape; the power of the erupting volcano was captured on canvas; and cameras took the desolate, rugged terrain of the American West and converted it into both a commodity and a comfort. What their images presented to Americans was the primary crucible in which the nation was being made, materially and in spirit.[39]

Encounters with the American environment, whether it was Edenic, malignant, or sublime, nearly always involved a relation with the divine, but not necessarily with a transcendent Christian God. After Hume and Kant, it was hard to argue for a natural theology, and after Darwin, harder still to celebrate it. In a context in which it seemed philosophically naive to read from living nature to God, a common move was to shift the locus of the sacred to living nature itself. For Alexander von Humboldt, the German naturalist and explorer, contemplation of the intricate yet global matrix of life on Earth, which connected the bird singing in a garden to the presence of a particular species of plant and onward to the chemistry of the local soil, geological formations, and systems of climate and weather, was an exercise in spiritual communion that required no thought of a creator.[40] For a generation of American explorer-scientists inspired by Humboldt, the accumulation of knowledge of the natural world offered the promise of a kind of completion, a comprehension, through immersion, of the unity in the midst of diversity that qualified the planet as a cosmos.[41] Still, as Henry David Thoreau observed, no journey to a far meridian was required for such intuitions; mind could meet the spirit in living nature just as readily on a walk near Walden Pond as in a survey of the western mountains. "Our voyaging," he wrote, "is only great-circle sailing."[42] The point of exploration was exploration of the self, which had to begin with a realization that man and the natural world were part of the same process of being. This was a metaphysics more easily perceived and expressed through experience than through doctrine or ritual; in principle, then, it was accessible to all. Nature religion, with its conviction that whatever divinity abided in man was sourced from and shared with every other element of life on Earth, has maintained a significant presence in the American spiritual marketplace ever since.[43] The epiphanies of unity and belonging that attended the perspectives made available by the space age—the views of the whole earth, the sense of man's incorporation into the universe—shared in this tradition even as they added something radical and new.

Aviation and Religious Experience

On a July morning in 1844, in the woods near Concord, Massachusetts, not far from Walden Pond, Thoreau's neighbor Nathaniel Hawthorne sat down to record his sense impressions of the natural world around him—the shapes of the trees, the play of sunlight and shadow, the sound and movement of squirrels, birds, and insects. Interrupting this quiet reverie, "into the midst of our slumbrous peace," there arrived the whistle of a locomotive, "the long shriek, harsh, above all other harshness, for the space of a mile cannot mollify it into harmony."[44] To the cultural tradition that identified America as a garden, in which a close attendance on nature would cultivate also order, virtue, and spiritual contentment, the sound of the locomotive presented a powerful challenge.[45] It signaled the birth of a symbol that became a frequent feature in American literature over the course of the nineteenth and twentieth centuries: the serenity of a natural setting shattered by the entrance of a machine.[46]

Yet technology was not always the antagonist of those who sought an opening to the divine. After all, as Walt Whitman attested, to watch a locomotive rolling by and to hear the reverberation of its shrieks across the hills was to encounter a form of sublimity: "Type of the modern! emblem of motion and power! pulse of the continent!"[47] In Paris in 1900, at the Great Exposition, Henry Adams stood before a forty-foot dynamo and "lost his arithmetic" in an effort to comprehend what it could do. "The economies," he observed, "were absolute, supersensual, occult; incapable of expression in horse-power." And so "before the end, one began to pray to it; inherited instinct taught the natural expression of man before silent and infinite force."[48]

Adams was accompanied to the exposition by Samuel Langley, appropriately enough, for it was the machine that Langley was trying (and failing) to invent—a craft heavier than air capable of powered and piloted flight—that promised, more than any other technology, to open new channels to religious experience. The first generations of airplanes could not travel to the altitudes reached by balloonists, but it was not long before they were capable of rising above layers of clouds to the upper troposphere and lower stratosphere. Here, the aviator was entering what Margret Dreikhausen has referred to as "a meditative space," where the sensation of motion abated and where, in the absence of atmospheric diffusion, there was a dramatic contrast between the intensity of the sunlight and the darkness of the sky.[49] At times the aesthetic effects of the experience were so distinct from those available on the ground that the aviator felt as if he had crossed a fateful threshold, releasing him completely from his ties to the Earth. That impression, declared Charles Lindbergh, was particularly profound at night, when "it seems that the world has rushed along on its orbit, leaving you alone, flying above a forgotten cloud bank, somewhere in the solitude of interstellar space."[50] In the 1950s, with the advent of jet aircraft that routinely ascended to the upper troposphere, clinical research pointed to the existence of a "break-off phenomenon," experienced by about 35 percent of military pilots. It was characterized "as a feeling of being isolated, detached, or separated physically from the earth." The pilots "perceived themselves as somehow losing their connection with the world."[51] In 1957, venturing well into the stratosphere in the course of an experimental balloon flight, David A. Simons gazed at a solid shelf of cloud below him and recorded his thoughts on tape: "It gives one a feeling of being in heaven, above the rest of the world where you can look down over the edge and see the poor, faltering mortals. It's a strange sensation: a quiet world, peaceful, bright and dark at the same time."[52] It was not simply that the airman had ascended toward the symbolic seat of the divine. At these high altitudes, in the quality of light and sound and speed, he might actually feel the whispers of the sacred.

The most celebrated evocation of the spiritual promise of aviation is contained in the closing lines of John Gillespie Magee's poem "High Flight":

> I've topped the wind-swept heights with easy grace,
> Where never lark, or even eagle flew—
> And, while with silent, lifting mind I've trod
> The high untrespassed sanctity of space,
> Put out my hand and touched the face of God.[53]

Magee presents a picture of himself aloft in Dreikhausen's meditative space, imagining some form of encounter with the divine. But what precedes this passage, comprising the main body of the poem, is an effervescent description of a joyride through the skies. Magee has danced "on laughter-silvered wings," "joined the tumbling mirth / Of sun-split clouds," and "flung / My eager craft through footless halls of air . . ." Little wonder that the poem swiftly became a favorite of military aviators, for it revels in the experience of another kind of communion—between the pilot and his plane—that was accessible only to those with the necessary natural skill.[54] Magee has "done a hundred things / You have not dreamed of" because he commands his "eager craft" with an "easy grace," akin to that of the practiced huntsman on his horse.

The aviation age, therefore, produced a striking analogue of the Thoreauvian image of man living and thinking in occult and instinctual synergy with nature. A pilot of any accomplishment did not stand before an aircraft and feel oppressed by its power; he climbed inside and sought to make that power his own. It is normal to read in aviation memoirs that flying came "naturally," as if pilot and plane shared will and sinew in common. "I have sensed," wrote Charles Lindbergh, "the harmony of muscle, mind, and mechanism which gives the illusion of life to substance until levers move with thought as hand or foot, until the rhythm of an engine is geared to the beat of one's own heart, and wing in turning flight seems an extension of one's own body."[55] For the Italian Futurists, the fusion of human flesh with the hard metal and propulsive force of the flying machine promised liberation not just from gravity but also from mortality: by such means man might finally secure the attributes of a god.[56] Futurism envisaged what we would now call a cyborg, but for most aeronautical engineers that remained at best a very distant dream, unlikely to reward an investment of time and resources anytime soon. They directed their efforts instead toward a cybernetics that perfected the interface of information and control between the pilot and his aircraft.[57] And it was precisely such technology that, for the best aviators in the most advanced planes, preserved the sense of flying "naturally." It was as if they existed in the same intuitive field. For Frank Borman, an air force research pilot, then NASA astronaut, "the wonderful, almost indescribable feeling created by being in perfect harmony with your machine" was unique to the experience of flying, whether in the skies or in space.[58]

Above the clouds or simply flying alone, pilots were susceptible to a cognitive reordering, making new moral distinctions between the air and ground, charging the confines and mechanics of their craft with presence and meaning. On a night flight from St. Louis to Chicago, Lindbergh looked down on the "phosphorescent moss" of human society below and wondered what would happen if he resolved not to return to it. How far could an airplane take him before he was forced to land? In the course of these ruminations he realized that in a properly adapted modern aircraft it would be feasible to fly nonstop between New York and Paris. It was a "vision," he wrote, "born of a night and altitude and moonlight."[59] Eight months later, two-thirds of the way across the Atlantic Ocean and exhausted after flying for more than twenty hours, Lindbergh experienced a different kind of vision: of ghostly presences inside his cabin and of himself losing substance, consorting with the phantoms in some immaterial realm. "The feeling of flesh is gone. I become independent of physical laws—of food, of shelter, of life. I'm almost one with these vaporlike forms behind me, less tangible than air, universal as aether."[60] Eventually, Lindbergh "re-formed slowly as a man again, returning from spatial distances to my plane and body, condensing and collapsing into earthly qualities."[61] But he was not the same man as before. The phantoms, he acknowledged, could easily be explained away as the projections of a fatigued mind, but he took them seriously nonetheless, just as he had the nighttime reverie that first set him on course for Paris. Thoughts and dreams were not simply derivations of the real; they played a creative role in the construction of a life. Lindbergh concluded that "vision and reality interchange, like energy and matter."[62]

For the airman, then, flight held the potential to stimulate new moods and insights in ways that were compatible with at least the more expansive definitions of religious experience. But the airman's epiphanies were usually his alone, occurring in conditions of solitude, in a single-pilot plane flying in an empty sky. For those on the ground, in the early heroic decades of aviation, the situation was different. At the airshow, above the trenches, or through the vigils that accompanied each new attempt to extend the airman's reach around the globe, flight was converted into the object of communal attention and concern. Thus, when observers expressed awe and apprehension, it was a response not simply to the aerial feats they were witnessing but also to their own position as members of a crowd. In 1909 Franz Kafka joined the multitudes attending the airshow at Brescia, and though he shared in the general excitement at the machines flying overhead, he also seems to have felt some portion of himself disappear. Louis Blériot, recent conqueror of the English Channel, took to the sky: "Everyone gazes up at him enraptured, in no one's heart is there room for anyone else . . . . We stand down below, quite left behind and insignificant and we watch this man." Exiting the show, Kafka and his companions "finally become independent beings once again."[63] A few years later much of Europe was looking at the airplane with the same sublime compound of reverence and dread. To the soldiers bogged down in the anonymous, intransigent slaughter of the western front, the "aces" wheeling in the skies above them almost appeared to be fighting in an entirely different war, one in which it was possible to move with grace and freedom and in which the nobility of martial endeavor was visible to all. From the perspective of the trenches, the "aces" could also seem like pitiless angels of death, delivering facile annihilations by means of strafing runs and bombs.[64]

And then, a decade on, the aviator returned as a redeemer. Major Pierre Weiss, the air force officer who had retrieved Charles Lindbergh from the crowd that enveloped him after his arrival at Le Bourget Field, commented later that Lindbergh's "miraculous appearance" had helped the French nation to rediscover its "desire for the infinite." Lindbergh, he declared, had "placed us in a state of grace."[65] Myron T. Herrick, the US ambassador in Paris, attested to "the electrical thrill which ran like some religious emotion through a whole vast population."[66] Back at home, Lindbergh's achievement was feted with a similar passion, which spoke of a need to compensate for recent disappointments and depletions. The flight yielded satisfactions that the Great War had not. Virtues that had seemed exhausted and provincial amid the racy metropolitan thrills of the Jazz Age and the material drives of a "business civilization" were now restored—in speeches, editorials, sermons, and poems—to the beating heart of national ideology. Certainly, it was noted, Lindbergh could not have accomplished what he had without modern engineering and effective planning and organization, but it had taken the sort of fortitude and independence synonymous with the American pioneers of old, at work in the wilderness, to actually face the odds and fly the plane to Paris.[67] And so the moment Lindbergh landed at Le Bourget, half a world away, America was touched too, as if by a better angel of its nature.

Though angels can bestow grace, they can fall from it too. By the mid-1940s many of the prophetic hopes once invested in aviation had been chastened by the ease with which the airplane had been assimilated within the aesthetics of fascism and more generally by the evidence of its lethal instrumentality in the Second World War.[68] Lindbergh himself, his reputation tarnished by his dinners with Goering and his leading role in the isolationist movement prior to Pearl Harbor, was not permitted to participate fully in the conflict when it came. Nevertheless, he saw enough of the destruction wrought by weapons of the air to chill his romance with the technology of flight.[69] Lindbergh turned back to the ground and turned into Thoreau, regularly venturing out into nature and wilderness to seek intuitive experience of an immortal "life stream" that connected a man to all men and to the environment of the earth and the cosmos beyond.[70] A decade earlier, Antoine de Saint-Exupéry had described a similar spiritual trajectory. In 1912 Saint-Exupéry had flown for the first time and called it a "baptism."[71] But his true conversion occurred later, in 1935, after he crashed his plane in the Libyan desert. Without food or water and weakening fast, Saint-Exupéry and his engineer wandered for four days, until they were saved by a Bedouin nomad who gave them water to drink. Here was the scene of conversion: from air to earth and from the solitude of flight to the community of man. For Saint-Exupéry, thereafter water was "a sensitive divinity": "Of all the riches in the world you are the greatest, and the most delicate, you who lie so pure in the womb of the earth." And of his Bedouin "saviour" he wrote: "You are our beloved brother. And I in my turn will recognize you in all men."[72]

After Guernica, Coventry, Dresden, and Hiroshima, the elations of flight no longer seemed so innocent. There was renewed interest, for example, in Sigmund Freud's 1910 discussion of Leonardo da Vinci, which attributed the artist's fascination with flying machines to an Oedipal mother fixation and a narcissistic desire for control over the world.[73] According to one clinical study of US Air Force pilots conducted during the war and published in 1952, the impulse to fly was closely connected to Oedipal fantasies of triumph over the father and freedom from women: "Every dangerous success is an indulgence of incestuous desires, at the same time that it defies and mocks the authority and power of the father and thereby brings reassurance of omnipotence and of the ability to withstand castration."[74] When John Magee reached out to touch the face of God, it seemed, it had not been a gesture of reverence at all, but a conscious and deliberate act of transgression.

Meanwhile, the interwar dream of a plane in every garage had dissolved, yielding to a new era of civil aviation dominated by large municipal airports, commercial airliners, and standardized air routes.[75] More people were flying, but they were sharing their air space with everybody else. They might still experience solitude, but it was not that of the airman in a reverie of separation from the earth; it was the isolation of the lonely crowd encapsulated in an aircraft cabin. Flight, disenchanted, was absorbed into the corpus of bourgeois social routines. The form of flight most closely attended by magical thinking—in which the single-piloted plane ascended in seclusion to an elevated altitude, aided by advanced technology that smoothed the transference of will and control between man and machine—was regularly experienced by only a few American aviators, those who belonged to the fraternity of military and NACA test pilots. However, with the arrival of the space age and the passage of these pilots into the astronaut corps and the ranks of national celebrity, whatever gift for epiphany they had developed in the course of their flying careers acquired a new significance. Such epiphanies, if replicated in Earth orbit or on the way to the moon, might potentially set the terms by which space travel was understood. Would the venture toward the stars be endorsed by Americans as more than simply a necessary Cold War countermove or a mode of Keynesian stimulus? Could they also come to invest in it a creditable hope of spiritual renewal?

Religious Experience in Spaceflight: Expectations and Limitations

In the early aviation age, those who speculated that flying would afford a sensation of being closer to God could reasonably determine, if they had a few dollars, to travel to a local airfield, hire a pilot and plane, and test the thesis for themselves. At the very least, like Kafka at Brescia, they could attend an airshow and personally apprehend the physical and moral force of the machines careering above them. In the space age, only a small, exclusive cadre of astronauts were permitted to ride rockets into the sky, and only in east-central Florida could Americans witness their departures at first hand. In the 1960s, however, it seemed plausible that modern technologies of communication would shrink the distances between the site of a national event and almost every family hearth in every locality in the country. The vigil for Lindbergh had been exciting; but now, via live television, there was the promise of immediate, real-time experience. In July 1969, shortly after *Apollo 11* was launched toward the moon, Richard Nixon asserted that television, even across the amplitudes of outer space, "brings the moment of discovery into our homes, and makes all of us participants." Indeed, Nixon declared 21 July, the scheduled date of the landing, a "National Day of Participation" and called on "all of our people, on that historic day, to join in prayer for the successful conclusion of Apollo 11's mission and the safe return of its crew."[76]

Certainly, a sense of corporate purpose pervaded the space program, a sense that a meaningful portion of the nation's hopes and aspirations had been placed in the solemn trust of the astronauts and those who were to help them rise toward the stars. Some of those aspirations were resolutely secular, but not all. In the early 1960s, Tom Wolfe has suggested, Americans regarded the astronaut atop his rocket as the Israelites had once regarded David as he advanced across the Valley of Elah to take on Goliath: his fate would be their fate and a test of the truth of their God against that of the heathen.[77] So the safe recovery of each astronaut, especially those who had performed their mission without conspicuous blemish, seemed to send a spiritual charge through much of the country, at least until safe recoveries began to appear routine. In addition, the assertion of Russian cosmonauts that their surveys of the heavens had empirically confirmed the official atheism of the Soviet system may well have tilted what Americans wanted from their spacemen: if not necessarily a discovery of the angelic host hovering in Earth orbit, then— please God—some minimal reassurance that centuries of Christian supplications directed toward the skies had not been actually wasted on a desolate vacuum, which no faith could survive.

Even as the Soviet challenge in space spluttered as the decade progressed, the anticipation that American astronauts might return from their adventures with a new intelligence of the sacred did not fade. The journeys they would make between the realms of Earth and heaven seemed almost shamanic.[78] "You get a feeling," asserted the CBS News correspondent Eric Sevareid just before the crew of *Apollo 11* left for the moon, "that people think of these men as not just superior men but different creatures. They are like people who have gone into the other world and have returned, and you sense they bear secrets that we will never entirely know, that they will never entirely be able to explain."[79] The astronauts were figures of existential significance not just because they were the country's champions in a fateful contest with a rival superpower but also because there was something about their journeys into space that spoke of a profound spiritual errand. Would they return to share new insights with those who had remained back on Earth, and what might their insights be? There were some standard conjectures.

By the early space age, astronomers had reached no final determination as to the size of the universe, but even their provisional calculations—in a range of billions of light-years—made a comedy of the Christian cosmological tradition. Yet it was possible still to look through a powerful telescope and maintain a pious eye. A divine intent could be intuited in the apparent order of the firmament, in its star systems and galaxies, and if one assumed the existence of an ordering mind, the endless magnitudes of space simply presented evidence of that mind's infinite power and reach. It seemed reasonable to suppose, therefore, that any astronaut who traveled beyond the murk of Earth's atmosphere would be exposed to such a brilliantly illuminated procession of stars, moving across a vista of such unfathomable depth, that he could not help but speak of his experience in terms of the sacred. "A number of clergymen," asserted *Time* magazine in a discussion of the religious implications of space exploration, "feel that growing knowledge of the immensity of the universe may prove to be a stimulus to renewed faith in God the creator."[80] But could not the astronaut just as easily gaze out from his craft into that immensity and, like Nietzsche's madman, apprehend "empty space breathing at us?"[81] Might he not return with the desolate news that the heavenly procession seemed without direction or purpose and that the earth itself was hopelessly adrift? Such thoughts, declared Paul Tillich in 1962, were disturbing to Christians and non-Christians alike: "They are crucial to every man's understanding of himself as a human being, placed upon this star in an unimaginably vast universe of stars."[82]

By the early 1960s, astronauts were not just going into space for a better view of the starry skies. They were preparing, at length, for an actual journey of discovery, to make landfall on the moon by the end of the decade. Both the mandating of a specific destination and the fact that it was the moon were significant. Since the Second World War, whenever Americans had imagined an encounter with other worlds, they had tended to assume that those worlds would come to them. Some, indeed, had declared that they had already been contacted, very often with the result that they had become prophets and apostles. Their contact experiences had convinced them that mankind was in dire need of a moral and spiritual reformation, and so—to effect that reformation, to evangelize the message they had been sent from the heavens—they had established their own churches and cults.[83] But now, beginning with the moon, mankind itself was dispatching emissaries into the cosmos, an enterprise that presented a more plausible prospect of cultural renewal. After all, it was frequently asserted, had not Columbus's report in March 1493 that he had discovered a new world awakened the people of a fractious and moribund western Europe to fresh, green horizons of social possibility; had it not prompted the revival of science, philosophy, and religious faith?[84] And though unlike America on the eve of Columbus's return, the moon was actually known to be there, and though it had surrendered much of its original mystery to the telescopic gaze, there survived in the culture some association with enchantment and some interest in its subtle effects on the earth—on the tides, on plant life, and just conceivably on the human personality.[85] Might the astronauts succumb to what remained of its spell? Or, perhaps worse, would their inspection tour of the moon complete the process of its desacralization, annul the best remaining metaphor for romance, and lock mankind more securely in the bleak, modern linguistic maze of computer code, corporate cant, and bureaucratic jargon?[86]

Or, alternatively still, the specific destination of the astronauts might prove less important to whatever new awareness they carried back with them to Earth than the general experience of being tested in the extreme environment of space. Possibly the point of the lunar excursions, as with the polar expeditions at the turn of the twentieth century, was not discovery but adventure.[87] It was not so much the prospect of an encounter with a new world that motivated the explorer and accounted for his fascinated public; rather, it was the prospect of an encounter with his own elemental self and the revelation of what capacities of faith, will, and judgment a man could retain in the most trying of physical conditions. The conviction that mankind needed to embrace the challenge of space in order to revive its self-confidence and develop to its full moral and spiritual potential animated much of what has been termed the *astrofuturist* tradition in the postwar United States, stretching from the science-fiction writings of David Lasser and Robert A. Heinlein through to Gerard K. O'Neill's detailed prospectus for human colonies in high Earth orbit.[88] In the spectacle of spaceflight could abide a vision of a better human future in the making. If that was the case, then the evolutionary role of the astronaut was to behave and to speak as if he was already ahead of the curve, to convince all Americans that they, like him, could come to feel physically and spiritually at home in the cosmos.

These hopes had a darker obverse, captured most compellingly in the short space-age fictions that J. G. Ballard started writing in the early 1960s: the apprehension that the astronauts would fail. They might, for example, inadvertently carry deadly microbes back with them from another planet, precipitating catastrophe at home; or they might become stranded in space, their supplies of oxygen running out, until at last their capsules were turned into tombs encircling the earth; or they might suffer a mental collapse, their "nightmare ramblings" disturbing "millions of television viewers around the world, as if the terrifying image of a man going mad in space had triggered off some long-buried innate releasing mechanism."[89] In scenarios such as these, cosmic imaginings would curdle. Far from lighting a path into the universe, the astronauts would have caused, instead, a fateful contraction of the human promise.

Indeed, much of what passed for "space science" in the 1950s and early 1960s took the form of anxious and often exotic speculations about how those commissioned to undertake spaceflight might respond psychologically to its particular demands and conditions. The capsules designed for NASA's original Mercury program, though they were intended to serve a civil rather than a military purpose, exemplified the trend in wartime and postwar weapons development toward complex "man-machine" systems. One corollary of that trend was a proliferation of connections between the field of psychology and the US military, for often, it seemed, the weakest link in these systems was the human operator.[90] Even the most stable of military aircraft might spiral to the ground in the hands of a pilot whose temperament made him basically ill-suited to flight or who had not undergone the psychological preconditioning appropriate to his task. So the military establishment welcomed the assistance of psychologists in both the selection and the training of its personnel. This involved, in many instances, creating profiles of the various specialized roles that service personnel might be required to perform, predicting the sources of mental strain that might attend each of these roles, and deciding which category of candidate, and what forms of preparation, would combine to best ensure that the man-machine system achieved and maintained optimal effectiveness. The government's formal adoption of the goal of manned space exploration opened another frontier for these applications of theories of mind. As officials of the newly minted space agency labored to develop a specification for the equally novel position of American astronaut, psychologists lined up to proffer them a stream of suppositions about the likely challenges that spaceflight would present to the mental stability of all but the most robust and well-conditioned of men.[91]

There was already an awareness, within emerging "space medicine" networks, of the break-off phenomenon reportedly experienced by many military pilots when they flew at high altitudes and attested to also by researchers who ascended into the stratosphere in a pressurized balloon.[92] Would not an astronaut, circling the earth beyond the outer edge of its atmosphere, be even more susceptible to such a phenomenon? Perhaps he would become so detached from the world that he would prove incapable of initiating the necessary procedures for return. But it was possible that the break-off phenomenon was simply one variant response to a more elemental psychological hazard: isolation. Given the limited payload capacity of existing rockets, it was likely that the first generation of American astronauts would be traveling into space alone, experiencing a distance from human society that had only the most forbidding of analogues in earthbound circumstance. Both the annals of polar exploration and the testimony of US prisoners of war held in seclusion during the recent Korean conflict suggested that profound social and sensory deprivation often resulted in psychosis.[93] Just a few years earlier, in a fresh account of his solo flight to Paris in the *Spirit of St. Louis*, Charles Lindbergh had revealed for the first time his encounter with "ghostly presences" over the Atlantic.[94] But psychologists could also draw on the evidence that was beginning to emerge from a range of controlled isolation experiments, some of which were expressly intended to replicate the conditions to be confronted by the astronauts—for example, through prolonged confinement in a space-capsule simulator.[95] Almost routinely, those who participated in these experiments started to hallucinate. "Deprived of sensory input," noted one clinician, "the mind is cut adrift and regresses inexorably into that Sargasso Sea of the primary process, where time disappears, where there is no order, quantity, direction, or rationality, where vivid, multicoloured hallucinations swirl and befuddle the senses." It was therefore vital, he asserted, that candidates for spaceflight be tested for their tolerance to social isolation and a withdrawal of sensory stimulus. Moreover, for the duration of an actual mission, the astronaut would have to be kept busy and stay in constant communication with the earth to preclude his mind's taking an inward, pathological turn.⁹⁶ What haunted these recommendations was indeed a scenario that might have been sketched by J. G. Ballard: the astronaut in orbit, commanded by voices and visions as if in a mystical trance, beyond the help of earthly powers but witnessed by all the world.

In their preparations for human spaceflight, then, NASA officials had to do more than master the technology and techniques that were to deliver an astronaut into orbit and thereafter return him alive to Earth, though that challenge by itself seemed formidable enough. They also had to ensure that the astronaut did not become so overwhelmed by the experience of traveling in space that he ceased to function effectively. Thus, the process by which the first seven astronauts were selected for the Mercury program revealed NASA's marked institutional preference for candidates whose personalities appeared to be well armored against sudden and spontaneous transformations. The candidate pool was restricted to military test pilots, that is, to a class of personnel who had already demonstrated an ability to withstand stresses at least comparable to those likely to be encountered in space. In psychiatric interviews and psychological tests, the candidates were probed for evidence of emotional instability and particularly for correlations with the Freudian hypothesis that a love of flight often had its origins in feelings of sexual inadequacy, reflecting a narcissistic drive or a wish for self-destruction. In addition, their mental resilience was evaluated through a series of stress experiments designed to simulate the conditions of spaceflight; among these was a sensory-deprivation test, with each candidate confined for three hours alone in a dark, soundproof room.⁹⁷ Psychological stability, noted an official review of the selection process, was "the most important consideration in evaluating a candidate." Physiological performance was of secondary importance.⁹⁸

The concern of clinicians that the human mind might become unmoored amid the vast and occult solitude of space continued to shape practices within the Mercury program for some time after the completion of the astronaut selection process. Even if NASA was confident that the process had identified the seven most stable and resilient candidates, it still strived to limit the number of environmental as well as operational unknowns with which they might be confronted in the course of their missions. Hence the repeated whirls around centrifuges to replicate the gravitational forces that would be encountered as their rocket pulled toward the sky; hence also the frequency of their excursions aboard jet transport planes, which, following a parabolic trajectory, offered its occupants a brief experience of weightlessness on each successive crest and dive; and hence too their many hours enclosed in mock-up Mercury capsules rehearsing procedures and flight plans.[99] Significantly, the NASA official charged with designing the training regime for the Mercury astronauts was a former navy psychologist, Dr. Robert Voas.

For the duration of each of the initial Mercury missions, from the time that they were woken up through prelaunch to splashdown and beyond, almost every move made by the astronauts, and almost every word spoken, was monitored for evidence of their mental condition. The first two Mercury astronauts, Alan Shepard and Gus Grissom, were examined by a psychiatrist on the morning of their flights, prior to putting on their pressure suits.[100] The first four astronauts—with John Glenn and Scott Carpenter following Shepard and Grissom into space—all received a thorough psychiatric assessment on their return to Earth.[101] During the flights themselves, consistent with clinical advice, a near-constant dialogue was maintained between the capsule and NASA's stations on the ground, while the astronauts sweated their way through a mission schedule larded with operational checks and reports, piloting maneuvers, and observations of Earth and sky.[102]

For most of the Mercury astronauts, no doubt, there was an easy, natural fit between their own personalities and the psychological type that NASA officials considered least likely to succumb to the stresses of spaceflight. However, it was also not hard to discern the value placed by the agency on emotional stability, and so, in the course of the selection process and afterward, during the competition for mission assignments, the astronauts were careful to conform to the ideal: reticent, rational, and stimulated more by the operational challenges of the space age than by the opportunities it seemed to present for markedly new varieties of experience. "I tried not to give the headshrinkers anything more than they were actually asking for," recalled Gus Grissom. "At least I played it cool and tried not to talk myself into a hole."[103] Over time, indeed, the psychologists who initially had been tasked with ensuring the mental soundness of the men NASA sent into space were themselves increasingly the ones who seemed to be unreliable. Their premonitions about the unraveling of the human psyche in conditions of isolation high above the earth had been advertised widely, but they received very little validation when the Mercury program eventually took wing. All six manned Mercury missions returned with their astronauts still competent and sane.[104] In 1962 NASA terminated its research into the psychological effects of spaceflight.[105]

In possession of the authority of mission experience, the Mercury astronauts were able to redefine the practice and purpose of their profession in a manner that not only rejected the salience of psychology but also broadly stigmatized any interest in matters of mind and sense as a hazardous distraction from operational needs. "If I dream," Wally Schirra told Oriana Fallaci, "if I get lost in wonder at the sight of a sunset, a color, I waste the flight and maybe my life."[106] It was an antiphilosophy, Fallaci concluded, that was very quickly internalized by most of those who followed the Mercury Seven into the ranks of NASA astronauts. "Lieutenant, what do you think of imagination?" she asked Roger Chaffee. Imagination, he answered, was necessary for the invention of machines: "But imagination must be held in check by a consideration of what is logical and useful, otherwise it becomes a childish instrument. And none of us are children."[107] Above all else, the response of these recent recruits to queries about the subjective experience of spaceflight or the larger meanings of the space age was policed by the memory of what had happened to the Mercury astronaut who had most conspicuously given attention to such themes, Scott Carpenter. Here was an astronaut who was not hostile to the psychologists, who willingly crammed his mission plan with observations and experiments, who greeted the first sensation of weightlessness with an exclamation of delight, who joyfully drank in the sunsets and the sight of the earth below, and who later described his flight as a "spiritual experience."[108] But in operational terms Carpenter's performance left much to be desired, though it was hardly catastrophic. He used too much fuel wheeling his capsule around reporting on the view, and reentering the atmosphere under manual control, he couldn't get his angles right and so overshot his projected landing point by 250 miles.[109] An astronaut was permitted to make errors, but not because he had become engrossed in nonoperational concerns. "That sonofabitch will never fly for me again!" declared NASA Flight Director Christopher Kraft, and Carpenter never did.[110]

The *Apollo 17* spacecraft sits atop its Saturn V rocket awaiting launch in November 1972. Before each Apollo mission, Cape Canaveral was illumined like a shrine at night. Anne Morrow Lindbergh wrote of the Saturn rocket that was to fire *Apollo 8* to the moon: "Radiating light over the heavens, it seems to be the focus of the world, as the Star of Bethlehem once was on another December night centuries ago."
Courtesy of NASA.

John Glenn and President John F. Kennedy are greeted by cheering crowds in Cocoa Beach following Glenn's flight aboard *Friendship 7* in February 1962. Senators told Glenn that his mission had been "a great medium for revival of spiritual feelings and spiritual values."
Courtesy of NASA.

"And God called the dry land Earth; and the gathering together of the waters called he Seas; and God saw that it was good." Earthrise as photographed during the mission of *Apollo 8*, prior to the crew's reading from Genesis, in December 1968. *Courtesy of NASA.*

Loretta Lee Fry presents NASA Administrator Thomas O. Paine with petitions totaling more than 500,000 signatures in support of the *Apollo 8* Genesis reading in March 1969. *Courtesy of NASA.*

The earth from space as seen by Russell Schweickart in March 1969. This photograph, taken by Schweickart from the porch of the *Apollo 9* lunar module, shows Dave Scott in the hatch of the command module against the backdrop of the curving earth. "I felt a part of everyone and everything sweeping past me below," Schweickart later attested.
Courtesy of NASA.

Wernher von Braun and the Saturn V rocket prior to the launch of *Apollo 11* in July 1969. Von Braun had experienced a religious conversion two decades earlier, just after his arrival in the United States. He declared that it was "profoundly important for religious reasons" that mankind travel to other worlds.
Courtesy of NASA.

President Richard Nixon greeting the *Apollo 11* astronauts in their mobile quarantine facility aboard the USS *Hornet* in July 1969. The president asked the *Hornet*'s chaplain, Lieutenant Commander John Pirrto, to read a prayer.
Courtesy of NASA.

The *Apollo 13* astronauts on the deck of the USS *Iwo Jima* after their recovery from the Pacific Ocean in April 1970. The ship's chaplain, Commander Philip Eldredge Jerauld, offers a prayer of thanks for their safe return. The plight of *Apollo 13* had prompted Congress to pass a resolution calling on all Americans to pray for the crew.
Courtesy of NASA.

Edgar Mitchell took this photograph of Alan Shepard as the two *Apollo 14* astronauts explored the lunar surface in February 1971. Mitchell recalled being "struck by an upwelling of obscure feelings" during his time on the moon. These feelings enlarged into a "grand epiphany" during his passage back to Earth.
Courtesy of NASA.

"I will lift up mine eyes unto the hills, from whence cometh my help." *Apollo 15*'s James Irwin recited this line from Psalm 121 as he and Dave Scott explored the Imbrium Basin, at the foot of the Apennine Mountains, in July 1971. Irwin, pictured here with Mount Hadley in the background, felt God's presence on the moon. *Courtesy of NASA.*

An artist's rendering of High Flight Lodge, intended to be the headquarters of James Irwin's High Flight Foundation. Irwin envisaged that the lodge, situated in the Colorado mountains, would serve as a center for High Flight's retreat ministry. The space-age design included "service modules" (offices, shops, lounges, and a restaurant) and "sleeping modules" as well as a chapel with additional outside seating, leisure facilities, and a space museum. The lodge was never built.
Courtesy of the Billy Graham Center Archives and High Flight Foundation.

Apollo 17 launches at night, December 1972. Isaac Asimov, watching from a cruise ship anchored offshore, noted that "it was useless to try to speak, for there was nothing to say."
Courtesy of NASA.

CHAPTER FOUR

Perhaps a Meaning to Us

The Apollo Missions as Religious Experience

Following Carpenter's flight, the astronaut who spoke with sincerity of a personal questing impulse or in florid anticipation of the poetics of traveling in space was the least likely to be trusted with an actual mission. And for those who did make it onto a rocket and up into the sky, NASA's institutional culture continued to exert a gravitational pull, discouraging the astronauts from offering much more than rote hosannas to the new sights they witnessed; their reports on new sensations tended to be dressed in self-conscious calm. But the deficits of excitement and awe in an astronaut's response might not have been a function of institutional culture alone. From inside a spacecraft, for example, it was frequently impossible to experience the full visual repertoire of the cosmos. The windows on the Mercury capsules may have afforded beautiful views of the earth and, during orbital flights, of the rising and setting sun, but they were dense and heavy, canceling the delicate luminescence of the stars in much the same way that the atmosphere does for observers on the ground.[1] Later, during the initial Apollo missions, astronauts complained that some of their windows were unusable because of frosting and condensation.[2] In the vicinity of both the earth and the moon, reflected light from the sun saturated the heavens and caused the irides of the astronauts' eyes to contract; only in midpassage between the two bodies or when traveling through the lunar shadow could they view the stars and constellations with real ease.[3] For long phases of their voyages, the Apollo crews were unable to see the earth or the moon at all because of the attitude and direction of their spacecraft.[4] Nor did excursions outside the capsule promise a reliable liberation from such constraints. Sometimes an astronaut's helmet visor misted up in the course of a spacewalk, rendering him nearly blind; and out on the lunar surface, the awkward bulk of his pressure suit could defy a desire to lean back, arc his neck, and gaze at Earth hanging high in the lunar sky.[5]

Even in those days and hours when the cosmos lay unconcealed, when the earth and the stars were processing past their window, or when they were actually traversing the open terrain of the moon, the astronauts rarely had much time to appreciate the view. They could not sit for a morning in the manner of Thoreau, slowly incubating epiphany. Inside the capsule there was an unending series of systems checks to perform, particularly when the mission hardware was new. Russell Schweickart flew on *Apollo 9*, the mission tasked with conducting the first operational test of the Apollo lunar module in tandem with the other components of the Apollo spacecraft system. He recalled that until the sixth day of the mission he hardly had a free moment to look out the window and observe the rolling Earth below: "Don't stop; you don't have time for that. And so out with the checklist and down through the day."[6] Outside the capsule, during spacewalks and moonwalks, the pace of work remained insistent. After laboring across twenty feet of open space to retrieve an experiment package from an inert Agena rocket, the *Gemini 10* astronaut Michael Collins imagined what he might have said about the experience: "'I found God outside my spacecraft.' Wrong, I didn't even have time to look for Him."[7] On the moon itself, conscious of public concerns about the expense of the Apollo missions, astronauts raced around seeking to maximize concrete scientific returns, setting up experiments, harvesting innumerable samples of lunar dust, rocks, and soil, the prompting voice of mission control always persistent in their ear. "Each minute had to count for something.... We kept our wonder in check, or at least under our breath," the *Apollo 14* astronaut Edgar Mitchell noted later.[8] During *Apollo 15*'s visit to the moon, David Scott and James Irwin drove themselves so hard that their heartbeats became irregular; subsequently, perhaps consequently, over the next twenty years Irwin suffered three heart attacks, the last of them fatal.[9] Only on the journey home was there the opportunity to rest, absorb, and reflect on all that had been experienced—at least in the case of the lunar-module pilot, whose major operational responsibilities disappeared when that module was jettisoned after the ascent from the lunar surface.[10]

Even when the earth, moon, and stars fully manifested themselves and the astronauts had the leisure to look out on the view, they did not always respond with appreciation, let alone with reverence. "It's just those damn Himalayas," commented the *Apollo 7* astronaut Walt Cunningham when asked what he could see outside his window.[11] The aesthetic spectacle without contrasted starkly with the material conditions of existence inside the cabin. Although the modular configuration of the Apollo spacecraft, unlike the Gemini capsules, permitted its crew to move around, it was still a bad place for a claustrophobe.[12] There was no prospect of personal privacy, while efforts to maintain a semblance of personal hygiene quickly proved futile. Without running water it was difficult to wash. The capsule carried only very limited supplies of replacement underwear.[13] Meanwhile, any food or drink that an astronaut allowed to escape, any gas that left his stomach, any urine, feces, or vomit that he did not discharge perfectly into the waste-disposal system or some other secure receptacle, and any dust collected on his pressure suit during his tramps across the moon might well continue to circulate around the cabin interior for the remainder of the mission. The odor that accumulated inside *Apollo 16* had become so noxious by the time the astronauts splashed down in the sea that the first swimmer to reach their capsule slammed its hatch back in their faces.[14] "*Life* magazine doesn't tell you about these things," observed the command-module pilot Ken Mattingly, but across the memoirs of Apollo veterans the spacecraft as a site of squalor and the unclean, uncomfortable body of the astronaut emerges as a rather common theme.[15] "I began to smell like a restroom, . . . It got so I couldn't stand my own company," attested Jim Irwin.[16] Grimy, rank, and acutely self-conscious—in such a state of being, was the astronaut a plausible subject for a hilltop experience, even before the most transcendent of panoramas? Was it possible, despite all this, for him to feel that he was bathed in grace?

The Epiphanies of the Astronauts

And yet, through all the pollution in their spacecraft, through their unrelenting attention to operational needs, through the acquired habits of emotional reticence that marked their corporate style, it was possible still for the astronauts to sense some movement of the spirit. Very often, their own accounts of their voyages attest to moments when they experienced a sudden new profundity of mood, a powerful feeling of reverence, melancholy, or joy. In most cases the mood was fugitive; it came and it went, leaving little apparent impression on the course of the astronaut's life and only the imprint of hackneyed piety on the surface of his prose. For a few, though, the mood seemed to be pregnant with meaning: it stimulated sustained reflection on the general order of existence, the latitudes of the human cosmos, and the obligations that bound the individual to the rest of mankind. In two instances at least, its effects manifested themselves at length in a striking change of vocation. By most modern definitions this category of mood must count as a religious experience. And what had caused such moods to first develop and deepen was an exposure to new sensory information—about the earth, the universe, and the moon and about the contingency of man amid the perfect indifference of infinity, in a physical environment without measure, direction, or time.

Their nominal objective may have been the exploration of space, but when America's early astronauts made visual observations, they were looking mostly at the earth. It was difficult, after all, to see much of the starry sky. The first two suborbital flights carried Shepard and Grissom from the east coast of Florida out into the Atlantic, over what Grissom termed "brilliant gradations of color" as the brown of the land yielded to the white sand of the shoreline, the green of the coastal shallows, and the deepening blue of the ocean sea. Grissom attested to his "sheer amazement" at the scenery below.[17] But these delights were not exclusive to a rocket ride into space; any aircraft traveling at an elevated altitude might have presented a comparable view. It was the orbital missions that really offered something new: a realization that the earth itself and not just the capsule, was in motion, and—every hour and a half—the sight of an accelerated sunset throwing all the colors of the spectrum across the horizon, followed by a sunrise forty minutes later. "Oh, look at that sun!" Scott Carpenter exclaimed as he watched it descend on the first of his three orbits. Then, shortly afterwards: "It's now nearly dark, and I can't believe where I am."[18]

In time, orbital flight came to seem like a routine accomplishment. Only a decade or so after John Glenn's first orbital mission, the third and final Skylab crew circled the earth more than twelve hundred times in the course of their eighty-four-day stay at the space station. Throughout that decade, America's astronauts had continued to take pleasure from gazing out of the window at the rolling Earth below or across to the horizon at the rising or setting sun. Aboard *Gemini 7*, Frank Borman and Jim Lovell witnessed their first dawn in space, "groping in vain to describe the awe we felt."[19] The experience of orbital flight, however, appeared to leave its most profound impression on those astronauts who ventured beyond the claustrophobic confines of their capsule, opening the hatch and hoisting themselves out into what was, suddenly, a sweeping visual field comprising the curving Earth and a bottomless black sky. With only an umbilical cord securing the astronaut to his spacecraft, it took no great poetic finesse to reach for metaphors of obstetrics. "That was the most natural feeling," Ed White declared to James McDivitt, his commander on *Gemini 4*, after White had completed the first "spacewalk" by an American astronaut. "Yeah. I know it," replied McDivitt, "You looked like you were in your mother's womb."[20] The metaphors don't quite work—was the astronaut intrauterine or had he actually been reborn?—but they are consistent with his complex apprehension of both freedom and dependence as, at a fierce velocity, he skimmed the edge of the world.

Standing on the porch of his lunar module during the Earth-orbital mission of *Apollo 9*, Russell Schweickart was unexpectedly afforded five minutes to register his position in the universe while his crewmate Dave Scott attended to a problem with his camera. "Now you're out there," he later recalled, "and there are no frames, there are no limits, there are no boundaries. You're really out there, going 25,000 miles an hour, ripping through space, a vacuum. And there's not a sound. There's a silence the depth of which you've never experienced before."[21] But the exhilaration of this encounter with infinity rather quickly seemed to yield to a renewed appreciation of the earth. By the sixth day of the mission, after most of its major objectives had been accomplished, the pace of work aboard *Apollo 9* began to ease, and Schweickart finally had the chance to gaze out of the spacecraft window. He looked first for the "friendly things," the cities of the American southwest: "And there's Houston, there's home, you know, and you look and sure enough there's the Astrodome—and you identify with that, it's an attachment." Gradually, with each passage around the world, the locus of attachment widened. Schweickart found himself "identifying with North Africa—you look forward to that, you anticipate it, and there it is." Eventually, by the end of the mission, his sense of connection had come to encompass the whole of the earth. "And somehow you recognize," he stated, "that you're a piece of this total life. And you're out there on that forefront and you have to bring that back somehow. And that becomes a rather special responsibility and it tells you something about your relationship with this thing we call life. So that's a change."[22]

Russell Schweickart, then, turned from the extremity of empty space back toward the earth and newly registered the latter as a cosmos, a system of life sustained on a single planet. It was to this planet, and not some starry futurity, that he now knew that he belonged, "a piece of this total life." Schweickart was perhaps unusual. Similar epiphanies of belonging were experienced by other Apollo astronauts, but for the most part only after they departed Earth orbit for the moon. On the journey out, as Earth shrank in size and the blackness around it grew, its blue and white colors became more brilliant. This was, wrote Michael Collins, "a sobering, almost melancholy, sight": the earth seemed beautiful and delicate precisely because from a distance it was clearly surrounded by annihilating space.[23] In the tradition of the sublime, it intimated human finitude and mortality. The lesson was reaffirmed by arrival in lunar orbit, for the moon appeared to be gray, cratered, and dead, a desolating vision of what could happen to a world. And so when their capsule circled back around to the scene of a lambent Earth rising above a barren lunar landscape, it was not uncommon for the astronauts to experience a powerful rush of emotion. "It was the most beautiful, heart-catching sight of my life," recalled Frank Borman, "one that sent a torrent of nostalgia, of sheer homesickness, surging through me."[24] For Borman, the voyage to the moon offered proof of man's dependence on God: the earth was a "miracle of creation," and everything else was "eternal cold."[25] In a more humanistic vein, Collins described the little blue sphere climbing over the craggy horizon as "home and voice for us": earthrise coincided with the restoration of radio contact with mission control, a reminder, if one was required, that here was the only source of comfort and aid available to the astronauts as they traveled across the heavens.[26] A little later, as the return journey began, Collins looked at the earth "in wonderment, suddenly aware of how its uniqueness is stamped in every atom of my body." But then, "I looked away for a moment and, poof, it was gone. I couldn't find it again without searching closely."[27]

When the earth summoned an affective response from the astronauts, therefore, it was as often as not because of what they now apprehended about the nature of the universe. Indeed, the turn toward "earth consciousness" could reasonably be considered a secondary phenomenon, a way of recovering a sense of center and ground after the primal shock of an immersion in the infinite. Astronauts who journeyed beyond Earth orbit rarely perceived outer space as continuous with the human cosmos. Only occasionally did they evoke—as John Glenn had earlier in the program—the role of an ordering hand that had set the earth and the planets and stars in their relations to one another, and even on those occasions Glenn's conviction was missing. "Someone, some being, some power," asserted Gene Cernan, "placed our little world, our Sun and our Moon where they are in the dark void." Yet he also acknowledged that the scheme "defies any attempt at logic"; it did not seem to be the product of a watchmaker God.[28] Moreover, as he stood on the lunar surface as the commander of *Apollo 17*, Cernan recognized the relativism of his situation. Earthly coordinates held no objective salience there. He remembered: "I had to stop and ask myself, 'Do you really know where you are in space and time and history?'"[29] Traveling back from the moon, the *Apollo 8* astronaut William Anders listened as a choral rendition of "O Holy Night" relayed from mission control became warped and distorted when the capsule's active radio antenna moved out of the line of sight with the earth, and he felt his own hitherto adamantine Catholic faith begin to buckle. If religious music could not survive an encounter with the universe, was there much hope for religious doctrine? And if the church could not claim to be universal, could it claim the authority to declare a transcendent law? "We're like ants on a log," Anders said. "How could any earth-centered religious ritual know what God's truth is?"[30]

A number of astronauts returned from deep space with a perception of its radical, unsettling otherness. A day after *Apollo 15* began its homeward journey, the command-module pilot Al Worden undertook the first-ever spacewalk outside of Earth orbit. Later, in a poem, he compared the experience to "Being let out / At night / For a swim / By Moby Dick."[31] On *Apollo 16*, Ken Mattingley made a similar excursion, wearing one of the helmets that earlier in the mission his colleagues had used on the moon. The helmet was equipped with a gold-plated visor designed to offer protection from the sun. Holding on to the handrail of the service module, Mattingley stopped to look around. The visor blanked out the light of the stars, Earth was not in view, and Mattingley had to turn if he wanted to see the moon. This was the only time in the mission when he truly understood how far away he was from home: "The entire world was within fifteen feet. And there was nothing else. That was a really powerful sensation. Never seen anything like it."[32] He was supported on the excursion by the lunar-module pilot, Charlie Duke, whose response to venturing out into "the empty blackness of space" would have interested researchers of the break-off phenomenon had such research not been terminated several years earlier. "The feeling of detachment I experienced was strange," Duke later observed; "it was almost euphoric, and I wondered what it would be like to float off into this blackness."[33]

For most of the astronauts, the Apollo-era navigations of deep space hardly felt like a consummation of the astrofuturist ideal. The universe offered no welcome to man, only an education in the negative sublime. There was one exception: the *Apollo 14* lunar-module pilot, Edgar Mitchell. Out on the lunar surface with the mission commander, Alan Shepard, Mitchell was "struck by an upwelling of obscure feelings," but it was only after they left the moon and started on the journey home that Mitchell had the time to think those feelings through.[34] As other astronauts had done, he gazed through a window at the earth and experienced a fierce sensation of belonging. There, "on that fragile little sphere," was "all I had ever known, all I had ever loved and hated, longed for, all that I once thought had ever been and ever would be."[35] Yet what was forged in this "grand epiphany" was not a new identification with the earth alone. Before joining NASA, Mitchell had studied at the Massachusetts Institute of Technology. He was well acquainted with both quantum theory and the recent empirical confirmation of big bang cosmology.[36] Now, in an "ecstasy of unity," as he coasted between moon and earth, he rapidly arrived at an understanding of what this cosmology really meant: that everything was connected. Whereas Collins had felt the earth in every atom of his flesh, Mitchell perceived the universe. "It occurred to me," he wrote, "that the molecules of my body and the molecules of the spacecraft itself were manufactured long ago in the furnace of one of the ancient stars that burned in the heavens about me."[37] And if every astronaut was made of the same stuff as the stars, then there should be nothing unnatural or disturbing about their journeys through outer space. For Mitchell, space travel was "an extension of the same universal process that evolved our molecules."[38] The human cosmos stretched into the infinite. From this "overwhelming sense of universal *connectedness*" he derived one further substantial insight.[39] It was an insight that could have been found in the natural philosophy of both Humboldt and Thoreau, not just in cislunar space, but it still impacted with decisive force on the trajectory of Mitchell's career. If everything was connected, having emerged from a common source, there was no fundamental divide between the realms of spirit and matter, religion and science. So if these realms were implicated in each other, what precisely was the nature of their relation? Mitchell was to spend the next twenty years of his life exploring answers to that question.

Both the sight of the distant earth and the passage through the vast primordial nocturne of outer space had the power to change not just the immediate mood of an astronaut but also, in some cases, his sensibility. The moon too could work a potent effect, in part because its position relative to the sun during the initial approach of the Apollo spacecraft often left it cast in shadow, a looming presence, in places pitch black or, where illuminated by earthshine, a spectral shade of blue gray.[40] "This cool magnificent sphere hangs there ominously," wrote Michael Collins, "issuing us no invitation to invade its domain."[41] Neil Armstrong described the view of the moon as *Apollo 11* flew through its shadow as the "most impressive" of the entire mission.[42] Then, as the craft eased down into lunar orbit, the scene changed. In the absence of atmosphere the surface of the moon, with its craters and peaks, could be seen with crystal clarity. For some of the astronauts this phase of orbital reconnaissance marked the end of their romance with the moon. From a height of sixty miles, the lunar terrain was not gentle, virginal, enigmatic. William Anders compared its appearance to that of "dirty beach sand."[43] To Collins, the moon was a "withered, sun-seared peach pit," of interest only to geologists.[44] But for those astronauts who were tasked with an actual landing, the moon retained an aesthetic appeal. Cruising over the lunar surface, they found a quality of grandeur in its austerity. "We could not pull our gaze away from the window," remembered David Scott, commander of *Apollo 15*.[45] There was, later attested *Apollo 16*'s Charlie Duke, "a beauty about this wasteland. It was spectacular and I was nearly overwhelmed emotionally."[46] This mode of appreciation carried over into the first Apollo moonwalks. The ceremonials of his initial footfall completed, Neil Armstrong looked around him, across the Sea of Tranquility, and declared that it had "a stark beauty all its own. It's like much of the high desert of the United States."[47] Buzz Aldrin stepped onto the moon shortly afterward, and observed, "Magnificent desolation."[48]

The early American moonwalkers were responding to regions of the lunar surface that had been selected as landing sites precisely because they were desolate and austere, lacking variety and visual drama. The lunar module had been tested in orbital flight, but the crew of *Apollo 11* would be the first to try to pilot it to the ground, and NASA was anxious not to combine this new operational challenge with the potential perils of a descent into hilly or mountainous terrain.[49] But after three successful landings, the operational unknowns had been reduced, and NASA scientists were pressing hard for the exploration of a wider range of geological formations. The remaining missions in the program ventured into more rugged and diverse areas of the moon: the highland plains near the Descartes crater (*Apollo 16*), the narrow Taurus-Littrow Valley, situated between steep, rocky walls (*Apollo 17*), and first, with the flight of *Apollo 15*, the eastern edge of the Imbrium Basin, at the foot of the Apennine Mountains, close to an ancient volcanic channel known as Hadley Rille.

It was here that the *Apollo 15* lunar-module pilot, James Irwin, had a powerful epiphany. The scenery was certainly impressive: Irwin's colleague, David Scott, compared the play of sunlight and shadow across the cratered landscape and the nearby mountains to that in Ansel Adams's iconic images of the American West.[50] But Irwin was particularly affected. He had spent his adolescence in Salt Lake City, from where he regularly trekked out to climb the Wasatch mountain range or, in winter, to ski through its many canyons and valleys.[51] Thereafter Irwin had always been most content in high, open places, during solo airplane flights, or in the hills and mountains around Colorado Springs, where he had lived with his wife, Mary, and their children before his selection as an astronaut carried them to Houston.[52] Indeed, the happiness of the Irwin family seemed to wither in the suffocating heat of the eastern Texas coastal plain, just as Irwin's own enthusiasm for space travel moldered quickly in the musty confines of the Apollo capsule during the outward journey to the moon.[53] But once he was in lunar orbit, Irwin's sludgy mood lifted as he gazed with fascination at the mountains of the moon. Here was nourishment for deprived eyes; here too was a promise of high, open country to explore. "I could see their beauty," Irwin recalled of the mountains, "and I felt I could almost reach down and touch them."[54]

Irwin and Scott landed at Hadley Base, and the next morning they walked on the moon. To Irwin, the landscape felt familiar, like his beloved mountains in Colorado or Sun Valley, Idaho.[55] It also felt providential. He had long associated the happiness he experienced at the high elevations on Earth, in the sky or among the hills, with a perception of proximity to the seat of the divine, and now, once more, he sensed God's presence around him. When Irwin had a problem erecting the power generator for the various scientific experiments that he and Scott were to leave on the moon, he prayed for guidance and immediately came up with a solution.[56] The next day, he spotted a strange, light-colored rock sitting on a base of gray stone, almost, Scott recalled, "as if it had been placed on a pedestal to be admired."[57] Scott wiped away some of the dust covering the rock and saw that it was composed of large, white crystals, an indication that it had once belonged to the moon's primordial crust. "I think we found what we came for," he told mission control.[58] Later the rock would be dated at more than four billion years old, close to the age of the solar system itself, and given the name Genesis Rock. To Irwin, the peculiar placement of the rock—"it seemed to say, 'here I am, take me'"—was evidence that its discovery had been the will of God.[59] He was convinced that man on the moon remained within the horizon of divine care and concern.

Heading back to the lunar module after this find, Irwin looked at the mountains surrounding Hadley Base and was inspired to ask Scott if they could hold a brief religious service. But Scott was not keen: it would require authorization from NASA headquarters, and in any case they already had a tightly packed schedule for the remainder of their stay on the moon. So the following day, as they returned to the lunar module for the final time, the sunlit mountains before them, Irwin offered up instead one of his favorite lines of scripture, from Psalm 121: "I will lift up mine eyes unto the hills, from whence cometh my help." He sensed, he later wrote, "the beginning of some sort of deep change taking place inside me," from the shallow, fitful religious faith that had marked his life before the moon to a new confidence in the power and agency of God. In contrast to his discomfort on the outward passage, Irwin felt "completely at home" on the journey back to Earth; he was sleeping much better, and his claustrophobia disappeared. After splashdown in the Pacific, Irwin reached over the side of the crew's liferaft and bathed his face in water; two months later, at Nassau Bay Baptist Church, he was baptized properly.[60] By this time, Irwin had begun to speak publicly about his conviction that God had been with him on the moon, giving him guidance.[61] In October 1971 he addressed an audience of some forty thousand Southern Baptists at the Houston Astrodome, after which invitations to offer his testimony at church services and other religious events started to pour in.[62] In the summer of 1972 Irwin brought forward his retirement from NASA because of a scandal surrounding the unauthorized carriage of four hundred commemorative envelopes aboard *Apollo 15* and their subsequent sale by the crew. At the same time, he moved to establish his own missionary foundation, to be named High Flight, after the poem by John Gillespie Magee.[63] He took his family back to the mountains of Colorado, where he planned to build a space-age spiritual retreat, complete with "sleeping modules" and "service modules," including restaurants, shops, recreation facilities, meeting rooms, a two-hundred-seat chapel, and a museum to contain memorabilia from Irwin's time as an Apollo astronaut.[64]

Any explanation of the epiphanies of the astronauts requires close attention to where they had actually traveled in space and what they had seen. Some sights and situations were more likely than others to elicit an affective or contemplative response. However, situational factors alone did not determine whether the experience of spaceflight would turn the mind of the astronaut to the deeper questions of human existence. David Scott had watched from the open hatch of the *Apollo 9* command module as Russell Schweickart, standing on the lunar-module porch a few meters away, absorbed the boundlessness of the universe; like Schweickart, he spent many hours during the second half of the mission looking at the earth.[65] Two years later he had traveled with James Irwin to the mountains of the moon. Though Scott had undoubtedly treasured these experiences, he did not feel that they had changed him. "Maybe I was transformed and I didn't notice," he commented.[66] Of those who had walked on the moon, he observed, it was the lunar-module pilots, not the mission commanders, who tended, in their subsequent careers, to make the more striking departures from the profession of astronautics. Perhaps this was evidence that they had also been more profoundly affected by their experiences in space, a circumstance that Scott thought attributable to the greater operational responsibilities borne by the commanders. Simply put, the lunar-module pilots had more time during the flight to let everything sink in.[67]

That might have been true, but not all lunar-module pilots were changed by the experience of spaceflight itself, and certainly not all emerged from that experience with a new appreciation of the sacred. For Buzz Aldrin, the first lunar landing represented the culminating achievement of his engineering career, but contrary to his hopes prior to the mission, it yielded no philosophy, no gleam of enlightenment, from which he could derive meaning and a sense of purpose when he fell back to Earth. A man who had walked on the moon, he thought, should have something profound to say to the world.[68] But Aldrin did not, and so, becoming increasingly depressed, he crawled deep inside a bottle. Indeed, the episode in Aldrin's life that seems closest in form to a religious conversion was not his tour around Tranquility Base but his decision nine years later to give up alcohol for good.[69] After *Apollo 12*, a sojourn aboard Skylab, and a period in the shuttle program, Alan Bean left NASA to become an artist, specializing in paintings that tried to express the sensory experience of visiting the moon. For an astronaut, this was certainly an unusual career move, but it was not a path ordained by any special epiphany on his lunar voyage. Out on the moon's surface, Bean had been too busy setting up experiments and collecting geological samples to really reflect on where he was.[70] Later, gazing up at the moon from Earth, he would wish for "some mystical feeling, but there never seems to be."[71] It is almost as if he became an artist in order to burnish and correct his own memory, to invest the Apollo expeditions with a romance and luminosity that, for most of the hard-driven astronauts, they could not have possessed at the time. Aboard *Apollo 16*, Charlie Duke exulted at the views of the earth, the moon, and the infinite depths of the universe, but he did not consider the experience religious. Indeed, in the immediate wake of his mission he was inclined to think that man's achievements in the space age had rendered God unnecessary.[72] When, in the late seventies, Duke became a born-again Christian, it was a transformation catalyzed not by his encounter with the cosmos but by his failure to find a fulfilling post-Apollo career, by the example of his wife's conversion a couple of years earlier, and by a Bible study course that had presented him with a choice: did he believe or disbelieve Christ's claim to be the Son of God?[73] Duke made his decision and then remade his life. His subsequent "walk with God," he said, "was more exciting than the walk on the moon had ever been."[74]

Of the eleven Apollo lunar-module pilots (two of whom flew without a lunar module), four—Anders, Schweickart, Mitchell, and Irwin—returned from their spaceflights with the sort of new information that compelled a reshaping of their structure of thought. Although the four acquired different information, with divergent implications, the fact that all were affected seemed to issue from correlations in mentality that predated their missions. Questions of God and mind had mattered to these astronauts even before they traveled into space, and yet they were also questions that had not been fully settled. Bill Anders was a thoughtful Catholic in the era of the Second Vatican Council and the encyclical *Humanae Vitae*, which opened the authority of the church to new qualifications and disavowals.[75] Rusty Schweickart, almost uniquely for an astronaut, was outspoken in his enthusiasm for progressive social causes. He had a background in astronomical research as well as a broad interest in the liberal arts, and though not religious in any conventional sense, he had friends who were involved in the Californian counterculture, with its experiments in spirituality, consciousness, and experience.[76] Edgar Mitchell had been raised as a Southern Baptist, but he was passing into agnosticism by the early sixties, when he started his graduate work at MIT. What he wanted from his scientific studies, and later from his participation in the space program, was clues to a cosmological synthesis that could replace the lost faith of his childhood. "What I was interested in," he recalled, was "the origin of our existence, the grand question of how we came to be."[77] So Mitchell remained alert for furtive evidence of a larger, unseen order, for patterns that seemed to transcend the normal distinctions between matter and thought, and in particular for signs of implicit communication across material distances and divides. In its preparations for the flight of *Apollo 14*, he observed, NASA itself "seemed to function with a common mind," as if every individual working for the agency "had been absorbed into the anatomy of a larger organism."[78] Mitchell was also planning to conduct a secret experiment in parapsychology on his journeys to and from the moon, seeking to transmit his silent thoughts to four collaborators back on Earth: if this could be done from outer space, it could be done anywhere.[79] Finally, Jim Irwin, who had been torn for a long time between secularity and the vestigial claims of faith, seemed in the months prior to his flight to be moving toward some kind of spiritual resolution. The potential hazards of the mission had served to concentrate his mind: "I wanted to be prepared in every way in case I didn't come back."[80] He joined a local Southern Baptist church, testified before the congregation about his relationship with Christ, and scheduled his baptism for after he returned from the moon.[81] But he could not yet attest to the sort of confirming experience of God's presence that Southern Baptists, more than any other major denomination, valued most highly.[82] What happened to Irwin on the moon, therefore, was just what he needed to truly belong in the church he had chosen back on Earth. Sometimes "the seeing eye" saw precisely what it had been looking for.

Astronauts and the Communication of Spiritual Experience

Not uncommonly, those who testify to an encounter with the sacred simultaneously declare an inability to describe that encounter in its fullest dimensions and depth. Some may acknowledge limits to their personal powers of articulation, but many class the problem as actually constitutive of religious experience itself. After all, if the sacred did not exceed what could be captured by human language, could it really be transcendental? The mystic, noted William James, is often skeptical of the efforts of others to detail and analyze mystical states of feeling.[83] Yet, as Rudolf Otto observed, experiences of the ineffable have rarely caused their subjects to adopt regimes of "unbroken silence"; indeed, "what has generally been a characteristic of the mystics is their copious eloquence."[84] For Otto, though the "numinous" state of mind defied strict definition, it could and should be discussed, for it was only through careful guidance and attention that a man could come to a consciousness of the numinous within him.[85] To share the details of a religious experience is also to invite others to assess whether that experience has any commonalties with their own; it may be the elemental act in the aggregation of a church.

Silence, moreover, was not an option for the astronauts. In formal contractual terms, following a deal originally negotiated on behalf of the Mercury Seven, they were obliged to talk to *Life* magazine, but that hardly exhausted the nation's sense of entitlement to their stories. Americans had paid many millions of dollars to send them out into space; yet victory in the race to the moon, when it was achieved, offered fewer geopolitical satisfactions than President Kennedy might have predicted in 1961, and it proved difficult to enthuse wide sectors of the lay public by itemizing technological spin-offs and discoursing on lunar geology. The capacity of the astronauts to convey the experience of spaceflight and to speak with conviction about its existential returns was almost all that stood between NASA and the popular perception that a lavishly funded rocket age had turned into something of a dud. Even after the Apollo missions ended, there continued to be a market for such testimonies, particularly from those astronauts who had actually set foot on the moon. A reasonable income could be earned from recycling anecdotes and reflections in memoirs, media interviews, after-dinner speeches, and other public appearances. Apollo veterans who tried to stay away from that circuit, or who preferred to offer their audiences something different, such as an induction into contemporary astronautics, could become objects of resentment.[86] "Tomorrow," wrote the *New York Times* on the eve of the first lunar landing, "we should know how, in fact, it feels to walk the moon."[87] Forty years on, it was an experience that the Apollo astronauts were still being asked to describe and explain.

Sometimes the expectation had the weight of a curse. For many of the astronauts, for most of their engineering and aeronautical careers language had primarily been a tool for the transfer of essential instrumental data. They spoke in acronyms and jargon because this seemed more efficient, at least within the discursive community of NASA, where such terms were understood. But then they were launched into space, and, with a hot mike to a listening world, they were required to convey an aesthetic and emotional response to an array of remarkable vistas. Lacking poetic facility and a deep reservoir of descriptive language to draw from, the astronauts, with an affable self-consciousness, returned again and again to the same dead adjectives. As he ascended Stone Mountain, above the Descartes highlands, Charlie Duke ruefully observed: "All I can say is spectacular, and I know ya'll are sick of that word, but my vocabulary is so limited."[88] When an astronaut did conceive of a fresh and poignant image—such as Bill Anders's description, soon after his return, of the distant earth hanging in the sky like a small, fragile Christmas tree ornament—that image was swiftly appropriated by his colleagues and reprised in memoir after memoir until it became a cliché.[89] And even the most generous and inventive of poets might find his accounts of his experiences drained of vitality and enthusiasm if he was asked to repeat them almost every day. Michael Collins declared: "If one more fat cigar smoker blows smoke in my face and yells at me, 'What was it really like up there?' I think I may bury my fist in his flabby gut; I have *had it*, with the same question over and over again."[90] It was a question for which Pete Conrad, the commander of *Apollo 12*, patented his own cheerfully deflective response: "Super! Really enjoyed it."[91]

In *The Spirit of St. Louis*, his second, Pulitzer Prize–winning account of his journey from New York to Paris, Charles Lindbergh sought to dissolve some of the distance between his experience in the cockpit and that of his readers by writing throughout in the historical present tense. By such a device, perhaps, the reader became a more immediate witness to Lindbergh's actions, thoughts, and feelings, including his drifts in and out of reminiscence, the deepening fog of his mental exhaustion, his cavorts with a sociable fraternity of ghosts, and his powerful sense of resurrection on reaching the coast of Ireland and seeing the people in the villages below waving at his plane. "Here are human beings. Here's a human welcome," he recorded. "I've been to eternity and back. I know how the dead would feel to live again."[92] Across the whole corpus of astronauts' memoirs nothing quite matched this.[93] Collins's *Carrying the Fire* was the best of the lot. It seems to have been written with Lindbergh's literary example in mind: the chapters describing Collins's two missions in space—aboard *Gemini 10* and *Apollo 11*—deployed the historical present tense, and Lindbergh contributed a foreword. He likened Collins's isolation in lunar orbit while Armstrong and Aldrin explored the moon to his own condition over the Atlantic, "when I seemed so disconnected from the world, my plane, my mind and heartbeat that they were completely unessential to my new existence." Lindbergh informed the reader that it was "easy to so identify" with the story told in *Carrying the Fire* "that you become an astronaut yourself."[94] Collins was too careful and circumspect to advance such a claim on his own behalf. There was a certain magic in spaceflight, he agreed, but it was hard to convey in words. He had enjoyed his solitude on the far side of the moon, but he had never been as remote from human contact as Lindbergh, for when he had come back around after forty-eight minutes, the radio connection with Houston had crackled once more into life. And there is no evidence in his memoir of a profound epiphany akin to Lindbergh's experience of discorporation.[95] Collins departed the space program with a mixture of deep satisfaction at what he had witnessed and achieved and "a mild melancholy about future possibilities."[96] He also had a much surer appreciation of both the earth's fragility and its position in Copernican space, "but basically I am the same guy. My wife confirms the fact."[97]

Still, of the four Apollo astronauts who returned from their flights with new intuitions of the sacred and their own relation to it, three made sustained efforts to communicate their experience to audiences at large and also to draw those audiences into shared explorations of its meaning. The exception was Bill Anders, who, through all the sermons and accolades that followed the voyage of *Apollo 8* and its crew's reading from Genesis while orbiting the moon, maintained a public silence about the dwindling of his own personal faith. He did so perhaps out of a sense of privacy, perhaps also because he was aware that the Apollo program would gain nothing, and indeed might lose a great deal, if Americans began to think that deep space made atheists out of the men sent to explore it.[98] In 1969 Anders took up a position in the Nixon White House, serving as executive secretary of the National Aeronautics and Space Council. There, even more than in Houston, it would have been evident that the well of public support for manned space exploration was in danger of running dry. Why risk saying something that could conceivably empty it for good?

Russell Schweickart, meanwhile, kept quiet about his epiphany for as long as he maintained some hope that he would be assigned to another mission. Even before his flight aboard *Apollo 9*, Schweickart was regarded with ambivalence by a number of his fellow astronauts, for they were adherents to the purist code of the "right stuff." Schweickart, with his interest in politics and culture, seemed to be implying that the code was insufficient, its purism provincial.⁹⁹ Then, on the third day of the mission, Schweickart succumbed to space-adaptation sickness, vomiting twice. He was not the first astronaut to become ill in space: Frank Borman had thrown up during the flight of *Apollo 8*. However, Schweickart's illness, unlike Borman's, forced a very public retrenchment of mission objectives. During his spacewalk the following day, Schweickart ventured no further than the porch of the lunar module, a much less ambitious excursion than NASA had originally planned. Also unlike Borman, Schweickart readily identified weightlessness as the cause of his condition and on his return to Earth volunteered to undergo an extensive series of medical tests to investigate why he had been so affected. Once again, Schweickart had failed to conform to fly-boy convention. He began his descent down the ranking list of flight-ready astronauts.¹⁰⁰

This would not have been a good moment for Schweickart to disclose his new and profound sense of attachment to the earth, for that would only have enforced the suspicion, already ambient within the astronaut office, that he was something of a flake. In any case, Schweickart attests, it took a number of years for the epiphany of belonging that he had experienced in orbit to "ferment and bubble," and only then was he able to articulate its significance.¹⁰¹ But by the conclusion of the Apollo program he was ready to speak, perhaps because in the prime crew assignments for Skylab he had been overlooked once again. At the end of 1972 Schweickart told *Time* magazine that in the course of his spacewalk "I completely lost my identity as an American astronaut. I felt a part of everyone and everything sweeping past me below."¹⁰² In the summer of 1974, having taken up a management position at NASA headquarters, he accepted an invitation to address the conference of the Lindisfarne Association, founded by William Irwin Thompson to study the role that the humanities, in tandem with the "new religions" of the East, could play in fashioning alternatives to a technological civilization that Thompson judged to be in an advanced entropic condition.¹⁰³ The theme of the conference was planetary culture, which Thompson predicted would quicken into being as man explored the anomalies in his existing categories of understanding, moving beyond those categories and their exclusions toward the sort of synthetic perspective denoted by the views of the earth made accessible by the space age.¹⁰⁴

Schweickart's presence at the Lindisfarne conference itself embodied this effort to start with an apparent anomaly and work toward the transcendent. Here he was, he wryly observed, "a nice, down-to-earth astronaut" speaking to "a far-out group like this" in order "somehow to share an experience which man has now had." What he had understood in the course of his epiphany was that he had made his voyage into space as "a piece of this total life," as "the sensing agent for man," with a special responsibility to communicate his experience to any audience that would listen. And so, seemingly seeking to erase the five years that had passed since his flight and to draw the imaginations of those at the conference aloft into the environment of space, Schweickart spoke in the present tense and the second person, referring throughout to an intuition of planetary unity that was happening now, to "you": "because it's not me, it's not Dave Scott, it's not Dick Gordon, Pete Conrad, John Glenn—it's you, it's we. It's life that's had that experience."[105] Schweickart later claimed that before beginning his talk he had had little idea what he would say, and that his words had seemed to flow spontaneously from some hitherto untapped well of experience and understanding, in the style of a charismatic preacher channeling the inspiration of the Lord.[106] It was, recalled one member of the audience, like listening to "a long, pauseless prayer," delivered by a man who seemed to be "amazed at what he was saying."[107]

Published in 1977, Schweickart's Lindisfarne address became an oft-cited and oft-reprinted text in the pamphlets, journals, and anthologies of the American peace and environmental movements.[108] Recorded and widely distributed as an audio cassette, it inspired documentary films, performance dance projects, and choral works.[109] Schweickart continued to speak about his epiphany to a variety of audiences, and in 1985, in collaboration with three Soviet cosmonauts, he founded the Association of Space Explorers, which sought to transcend Cold War divides with the affirmation that whatever the differences in the governing ideologies that propelled their missions, those who traveled into space always returned with the same, shared reverential perspective toward their home planet.[110] But Schweickart, whose greatest moment of spiritual adventure occurred on the porch of the lunar module during its circuits around the earth, lacked the aura that seemed to surround his moonwalking peers. Even had he wanted to, then, it would have been difficult to convert his personal epiphany into a credo for some kind of church. He continued to need a day job, serving, for example, as chair of the California State Energy Commission.[111] He was ostracized by NASA and a few of his former Apollo colleagues for his attempts to engage in citizen diplomacy with the Soviet Union via the Association of Space Explorers.[112]

His epiphany may also have lost some of its prophetic power when Schweickart himself began to change his mind about its meaning. A year or so after his Lindisfarne address, Schweickart became an enthusiastic advocate of the proposals advanced by Gerard O'Neill for human colonies—akin to garden cities—suspended in cislunar space.[113] Perhaps the identity of mankind was not so synonymous with the earth after all. By the late 1990s he had reconceived his epiphany as a futuristic vision, not now an expression of longing for and belonging to a planetary home but instead a threshold moment, a backward glance, before a spacefaring civilization left for the stars. "That enabled me to really see for the first time the implications of what going into space really means. . . . We're off to other planets, we're off to other worlds, different atmospheres, different gravity, weightlessness, people, kids being born in weightlessness. Who knows what is coming? But that is a change in history, and we're fortunate enough to live at that moment."[114]

In contrast to Schweickart, Edgar Mitchell had no ambition for another flight. Realistically, in an era of retrenchment, where could NASA take him that would be better than the moon? What he wanted most to explore was the implications of the "grand epiphany" he had experienced on the return journey to Earth, his "overwhelming sense of universal *connectedness*."[115] It was unlikely that NASA would be a supportive home for such investigations. When, a few days after splashdown, the press revealed details of the secret experiment in parapsychology that he had conducted aboard *Apollo 14*, Mitchell was neither chastised nor encouraged by the agency's senior management. He detected in their lack of response "an undercurrent of disregard" for research of this kind, a view that it was not proper science.[116] In 1972 Mitchell resigned from NASA, and shortly thereafter he founded the Institute of Noetic Sciences (IoNS), envisaged as an organization that would foster the study of consciousness by cultivating networks of like-minded researchers and, with the assistance of a few sympathetic wealthy donors, channeling funds toward significant scientific projects that otherwise would be unlikely to attract support from mainstream grant-giving bodies.[117]

It seemed like a propitious moment for embarking on such a venture. The IoNS set up its offices in Palo Alto, California, proximate to all the virtuoso sects and syndicates of the San Francisco counterculture and also to the Stanford Research Institute, where Mitchell had found two scientists who were willing to conduct empirical studies of alleged psychic phenomena. Mitchell arranged for the scientists to test, under laboratory conditions, the claimed psychokinetic and telepathic powers of a young Israeli magician, Uri Geller. The announcement by the Stanford scientists that Geller had achieved positive results in their experiments—results that seemed to defy conventional scientific explanation—precipitated a wave of public interest in parapsychology and made Geller an international celebrity.[118] Defense and intelligence agencies took an interest.[119] More broadly, across the counterculture, in certain stations of academe and sections of the lay mainstream, there could be identified a shared and strengthening conviction that purposeful inquiry into the phenomena of mind presented the most likely source of a fresh transformation of scientific paradigm, of the kind that Thomas Kuhn had described a decade or so earlier in *The Structure of Scientific Revolutions*.[120] Although many skeptics might still cast as mystical and unscientific the claim that consciousness could work independent effects on objects of mass and matter, they were increasingly met with well-informed references to the field of quantum physics, with its discovery of invisible forces at play in the universe, and to recent historical studies asserting the continued influence of the occult and magical traditions into and on the founding era of modern science.[121] It seemed reasonable for Mitchell and his associates to propose a synthesis between an older, metaphysical heritage and the Einsteinian cosmos. If there was evidence that the physical world could be plastic to currents of human thought and will, it was because consciousness and matter were derived from a common source and continued to be connected through the common attribute of energy. Such a synthesis, moreover, might have transformative significance well beyond the realm of science. Mitchell was inspired by the mission of healing a troubled and divided planet by fostering awareness of the fundamental unities linking man with every man, and mankind with the earth and stars. "It is the only thing," he declared in 1974, "that will counteract contemporary crises and bring meaning, direction, and fulfilment to people."[122]

The reformation did not come. In the tight economic times of the mid-seventies, the IoNS struggled to do much more than provision for its own survival. It failed to attract the sort of donations that would have allowed it to fund a comprehensive program of scientific research into the phenomena of mind. In the absence of an extensive record of credible experimental results, in a field that was, as Mitchell himself observed, notorious for its frauds, the scientific establishment continued to be hostile, and its paradigms remained unchanged. Some within the institute felt that Mitchell's interest in laboratory work had yielded rather minimal returns. Moreover, they wanted the IoNS to explore a much broader range of spiritual questions, and not simply those that Mitchell judged explicable in terms of physical principles. The institute, they argued, should become more like a church, and Mitchell more like a priest. Mitchell resisted, and in 1982 he was ousted from the chairmanship.[123]

The IoNS continued to grow, but that growth was accompanied by a further diffusion of focus, a further attrition of intellectual rigor. After a few years, seeking to rebalance its activities, the IoNS board invited Mitchell to return.[124] Now, more than two decades later, the IoNS has moved to a two-hundred-acre campus north of San Francisco and claims a membership of close to thirty thousand.[125] Not many Apollo program spin-offs can attest to the same durability or breadth of constituency. But does the IoNS, in its current interests and operations, really embody the consummation of Mitchell's original ambition back in the early seventies, as he sought to make sense of the epiphany he had experienced on the journey home from the moon? For the most part, it has failed to persuade the mainstream scientific community to make a reckoning with consciousness. And the genial welcome that the IoNS still extends to almost any dreamy expression of faith in the creativity of mind and spirit seems rather at odds with Mitchell's own long, stringent effort to elaborate a "dyadic" model of reality that defined the interconnections between the material and the mental. Mitchell's memoir, *The Way of the Explorer*, is no easy bedside read: the second half of the book is essentially a treatise.[126] Andrew Smith has described an IoNS conference in Florida at which Mitchell delivered a lecture lasting four hours. At an event in the evening, Smith was dragooned into a communal "Dance of Universal Peace." Twirling in a circle, chanting "I am sunlight/I am shining," he noticed that Mitchell remained "anchored to his seat as if stapled to it by Zeus himself."[127]

It took Mitchell more than twenty years to fully comprehend the meaning of his experience in space. It took Jim Irwin about a month. After all, he had journeyed to the moon having recently made a decision for Christ and having arranged to be baptized following his return. And he had had no doubts about the reality of God's presence as he walked out on the lunar surface. Irwin knew what had happened to him; to what extent, though, did he want to share the experience? The enthusiastic response from the local churches to which he offered his initial testimony and then the reaction to his appearance before more than forty thousand Southern Baptists at the Houston Astrodome convinced him that his new mission was indeed to speak, to lend his celebrity as an astronaut to the purpose of commending God to the world.[128] Unlike Mitchell, who was challenging the resolute materialism of the natural sciences, Irwin was pushing at an open door. To the Southern Baptist Convention, his arrival was like a gift from heaven. Here was a genuine American hero ready to move out across the country and proclaim a simple Christian message. God had been there on the moon, Irwin would tell his audiences, because there were no limits to his power and love. God had a plan for everyone; all they needed was the faith to place themselves in his hands, as Irwin himself had done during his excursions around Hadley Base.[129] Soon Irwin was doing more than just giving witness; he started to invite audience members forward to make their own decision for Christ. It is the clearest example of an astronaut seeking to convert his own spiritual experience among the stars into the building of a church on Earth.[130]

Truly effective evangelicalism, however, demanded process and structure. Left to their own devices, the people Irwin delivered to God were likely to drift away again in time. In June 1972 Irwin explored the option of joining the Billy Graham Evangelistic Association but was advised by Graham that the possibilities presented by an astronaut ministry were "so unique" that he should form his own organization.[131] Shortly thereafter, Irwin announced the creation of the High Flight Foundation, combining Christian outreach with plans for a retreat center close to Colorado Springs.[132] But he also contracted much of his time to the Southern Baptist Convention. For a season, Irwin became a near-constant presence on the convention's radio and television programming, while the Foreign Mission Board made use of his worldwide fame to give new momentum to its work.[133] "We rejoice at the prospect of his assisting in several mission fields," wrote the FMB consultant Joseph B. Underwood to Bill Rittenhouse, pastor of Irwin's church in Nassau Bay, Texas, who was soon to become the executive director of High Flight, "and we believe that his testimony would make a tremendous impact for the glory of the Lord Jesus Christ."[134]

Throughout the rest of 1972 Irwin journeyed far and wide on behalf of the Foreign Mission Board, including a month-long tour of Asia, New Zealand, and Australia. In almost every country on his itinerary Irwin's visit attracted the attention of the national media and elicited invitations to meet with political leaders, including the prime minister of Japan, the presidents of South Korea, South Vietnam, and the Philippines, and Madame Chiang Kai-Shek in Taiwan.[135] He held a number of religious rallies, with the venues frequently filled to capacity. During one such event in Taipei, more than a hundred members of the audience responded to his call to come forward to accept Christ.[136] The FMB had committed considerable resources to Irwin's tour, but, Underwood reported, "the expense has been more than justified by the publicity, by the positive testimony given in so many places, reaching such great multitudes that otherwise are never contacted by the gospel of Christ."[137] Back in the United States, by early 1973 Irwin was being described as "the leading preacher on the independent evangelical circuit," with much of his schedule already booked for the remainder of the year.[138] In February Irwin announced his plans for an ambitious summer retreat program for the former prisoners of war recently released from Vietnam, together with their families.[139]

A few weeks later, in April, however, Irwin suffered a serious heart attack, and his outreach ministry, which had depended on his ability to fly himself around the country, was grounded for the next two years.[140] High Flight never achieved take-off velocity again. The POW retreat program did go ahead, but a promise from a Houston businessman to underwrite its costs was rescinded at the final moment, leaving Irwin and High Flight with heavy debts that took years to repay.[141] There would be no permanent space-age retreat center nestled amid the Colorado mountains. Irwin's health remained precarious, with another heart attack in 1977. He lacked the vigor and resources to take advantage of the bull market for evangelicalism at the end of the decade. Although never moralistic in the mode of a James Robison—his family life was perhaps still too antic for that—Irwin joined the march of Southern Baptists in the late 1970s toward doctrinal conservatism and a position of biblical inerrancy.[142] In 1978 the same astronaut who had discovered the four-billion-year-old Genesis Rock on the lunar surface and perceived in that discovery the workings of providence endorsed a fundamentalist text that argued that the moon, along with the earth, had been created by God no more than ten thousand years earlier.[143] Consistent with the creationist contention that the fossil record was actually the product of the Great Flood, not evidence of an ancient earth, Irwin traveled a number of times to Mount Ararat in search of Noah's Ark.[144] "As far as I'm concerned, Noah's ark did exist," he declared in 1984. "There's no question about it. If we could find evidence of Noah's ark and convince the scientific community of it, then maybe they'd rethink the origin of the earth."[145] He was looking in the high places once again for a message to bring back home. If Irwin came down the mountain with proof of the ark's existence, the result might be a global resurgence of faith—of the kind that he now knew he could not achieve through his ministry alone. Though Irwin's status as a former astronaut helped open up access to Ararat, located as it was close to the sensitive Turkish-Soviet border, no relic presented itself to him providentially in the manner of the Genesis Rock on the moon. By the time Irwin died in 1991, following another heart attack, he had faded to the margins of the modern evangelical scene.[146]

Spaceflight and Secondary Religious Experience

To the extent that spaceflight was a spiritual errand, undertaken by the astronauts on behalf of mankind, its returns were rather mixed. Some toured around the heavens but found nothing of spiritual interest there. Some, in their encounters with the earth, the moon, or the stars, registered a strange deepening of mood, but then the moment passed, never to be recovered or evoked with much success in writing or speech. A few did experience such moods as a source of new conviction, but the messages conceived in a splendid Apollonian isolation were required on return to compete in a highly complex spiritual marketplace, profuse with rival cosmologies. The astronauts may have traveled into space in the name of all mankind, but whatever insights they had accumulated there were unlikely to entirely transcend earthly religious particularism. Edgar Mitchell scrutinized quantum theory for clues to the divine. But for Jim Irwin, "there are things that God does not intend man to understand, things that man is to take on faith."[147] To the evangelical imagination, Mitchell's notion of divinity was esoteric and recessive; it was too implicit in the substance of nature to do the work of saving souls. Dotty Duke, wife of Charlie, observed: "It's not the same God."[148] This was the fate of spiritual lessons conceived in the open cosmos: back on Earth they seemed to resonate most clearly in the context of closed structures, within church walls.

Still, the capacity of the space age to occasion religious experience was not confined to the corps of astronauts and those transformed by their testimonies; and not every affective response registered on the ground was obviously reducible to some preexisting sectarian perspective. To the astronauts sitting high up in the nose of the Saturn V rocket, the moment of lift-off was signaled by modest vibration and a rumbling noise far below.[149] From the viewing stands at Cape Canaveral, in contrast, the raw power of the rocket was manifest in a fierce spectacle of flame and bone-shaking waves of sound, as if an impassioned deity were at work. The astronauts, of course, were also absent in space for whatever earthly rituals, vigils, and impressions of collective experience attended their missions. Quarantined upon return, the crew of *Apollo 11* viewed videotapes of the television broadcasts that had tracked their progress to the moon, including images from around the world of crowds gazing intently at screens as Armstrong stepped off the ladder onto the lunar surface. "Neil," Buzz Aldrin commented, "we missed the whole thing."[150] And what of the writers and poets who had observed the same scenes? Would they use their craft to transfigure the technical achievement of Apollo into something sublime or expressive of the sacred? Could they fashion a common sense of it, as the astronauts could not?

Seventy years earlier, at the Paris Exposition, Henry Adams had entered the hall of dynamos and wondered at their power. He thought of the cathedrals at Chartres and Amiens, built by faith in the Virgin. Here were "two kingdoms of force," one traditional, the other new, one moral, the other not. It troubled Adams that he did not know which was the greater.[151] The same question haunted those who witnessed the Apollo launches. From a distance of a few miles, on the night before each launch, Cape Canaveral had the appearance of a shrine. Illumined in white light, the Saturn V rocket nestled in the embrace of the mobile launcher platform like some machine nativity. "Radiating light over the heavens," observed Anne Morrow Lindbergh, who was present to attend the launch of *Apollo 8*, "it seems to be the focus of the world, as the Star of Bethlehem once was on another December night centuries ago. But what does it promise? What new world? What hope for mortal men struggling on earth?"[152]

Then, the next morning the analogues changed, changed back, and then changed again, like the efforts of Ishmael to compass the White Whale. At the launch of *Apollo 11*, Norman Mailer, just as Adams had, "lost his mathematics." He "never had to worry again about whether the experience would be appropriate to his measure." The best he could offer was metaphor: first, Milton's hellish cataracts of flame, thereafter followed by the phoenix. "Two mighty torches of flame like the wings of a yellow bird of fire flew over a field," and the rocket—"this slim angelic mysterious ship of stages"—slowly and silently started to rise, "white as a ghost, white as the white of Melville's Moby Dick, white as the shrine of the Madonna in half the churches of the world." And seconds later there was sound, "a crackling of wood twigs over the ridge," building to "a cacophony of barks," to "a thousand machine guns firing at once," to a thunder "conceivably louder than the loudest thunders he had ever heard," to "an apocalyptic fury of sound equal to some conception of the sound of your death in the roar of a drowning hour." The shape of the rocket in the sky was "like a thin and pointed witch's hat, and the flames from its base were the blazing eyes of the witch." Mailer heard himself, high-pitched, saying: "Oh, my God! oh, my God! oh, my God! oh, my God! oh, my God! oh, my God!"[153]

But was this a hymn of praise or an ululation at the scene of a killing? There were pieties in Mailer's imagery but also death and the devil and the powers of the occult. In the rampage of its launch and the sublime majesty of its ascent, the Saturn rocket seemed like either an instrument of the divine or a sign of its negation. "What you watch is a biblical scene," commented Eric Sevareid on the *CBS Evening News*: "There's a gratitude, a thanksgiving, really a reverential sensation as you watch this."[154] Reverend Ralph Abernathy, leader of the Southern Christian Leadership Conference's Poor People's Campaign, traveled to the Cape on the eve of the lunar landing mission to protest against the space program as "an inhuman priority" when there were still millions of Americans subsisting in conditions of poverty and hunger. NASA invited him to observe the launch from the VIP stands. Interviewed afterward, Abernathy admitted that as the Saturn V had lifted off from its pad, he had for some moments forgotten all about the poor: "I was one of the proudest Americans as I stood on this soil, on this spot. I think it's really holy ground."[155] The laureate of uber-rationalism, Ayn Rand, however, disdained these sorts of consecrations. "What we had seen, in naked essentials," she declared, "was the concretized abstraction of man's greatness." In the dimensions of the force generated and in the quality of the control exercised over it lay the ultimate confirmation that the world had no need of God.[156]

It was *Apollo 17*, the final manned mission to the moon, that served up the greatest launch spectacle of all, for it departed at night, the flames from the Saturn rocket flooding the horizon with light until it was high and distant in the Florida sky, when the darkness returned and the rocket disappeared into the canopy of stars. By that time, however, its power seemed less portentous. Few were claiming the launch for either God or godless man. In the three years that had passed since *Apollo 11* the assumption that NASA could simply retool its manned spaceflight capability and then set its sights for Mars had yielded to a mood of diminished expectations. Thus, when *Apollo 17* rose up from Cape Canaveral, it was not the seal of some new covenant but the last sensation of a show condemned to close. A cruise ship, the SS *Statendam*, sailed from New York for the launch. On the way, passengers were to attend a series of seminars led by writers and scientists on the theme of man's future in space. It proved a difficult sell: the majority of berths were not taken, and some of the scheduled speakers did not appear.[157] William Irwin Thompson was aboard, and he freely exulted in the launch, but he also perceived "a sunset-effect to this technology of rockets." The space program had returned mankind to "our lost cosmic orientation," but now that orientation could be developed in different, gentler ways. A few months later he founded the Lindisfarne Association.[158] Others on the cruise marveled at the pyrotechnic display but then fell into silence, as if they had detected some deficit of purpose in the rocket, some lack of conviction in themselves. Isaac Asimov observed that "it was useless to try to speak, for there was nothing to say." Standing behind Asimov, a young man offered a poor facsimile of Mailer's invocation upon the ascent of *Apollo 11*: " 'Oh, shit,' he said, as his head tilted slowly upward. And then, with his tenor voice rising over all the silent heads on board, he added, 'Oh, *shi-i-i-i-i-it*!' "[159]

For the majority of Americans, of course, the Saturn launches did not shake flesh and bone: they watched each launch on television. "You can see it that way," asserted Eric Sevareid, "you can hear much of it, but you can't feel it."[160] The visual experience was reduced in scale, frame, and intensity: the flames did not burn so brightly, the roar of the engines did not stampede. Doubtless, some affect was possible. "I identified with the rocket!" recalled the heiress and self-titled futurist Barbara Marx Hubbard, who watched the launch of *Apollo 11* from her mansion in Lakeville, New York. "I felt myself rising in space, breaking through the cocoon of the sky and moving into the universe." She continued: "I cried uncontrollably as it rose into space, the words 'freedom, freedom, freedom' pounding in my head. I was so embarrassed I had to leave the room."[161] But such projections were probably unusual. On television, observes David Nye, "all lift-offs looked pretty much the same."[162] It was difficult to tell the difference between the titanic power of the Saturn V and the candle wattage of the Atlas boosters that had lugged the Mercury astronauts into orbit only a few years earlier. For all but the most jaded of Floridians, a Saturn V launch witnessed at first hand was an experience that endured; on-screen the sensation weakened with each facile repetition. In John

Updike's novel *Rabbit Redux*, set in the decaying Rust Belt town of Brewer, Pennsylvania, during the summer of 1969, Harry "Rabbit" Angstrom and his father head to a dive bar after knocking off work. That morning, *Apollo 11* had been launched toward the moon, and the bar television, its sound turned off, replays the footage for the twentieth time, the Saturn rocket like a boiling kettle, lifting slowly, and then its "swift diminishment into a retreating speck." There is no contagion of transcendence: "The men dark along the bar murmur among themselves. They have not been lifted, they are left here."[163]

In the spectacle of launch, then, NASA afforded the nation one kind of encounter with the sublime. In their photographs of a distant, spectral Earth the lunar astronauts offered another, and here the distinction between the primary experience and the viewing of its representation seemed rather less categorical. Both were essentially soundless, and for the onlooking astronaut the original scene was already framed—by either the spacecraft window or the visor of his spacesuit. Moreover, man had been prolific in his imaginings and representations of the earth as a planetary sphere for centuries before the space age provided the means to look on the actuality and capture it with a camera.[164] There was wide anticipation of such perspectives prior to the lunar missions that finally made them possible, anticipation that was often charged by some prediction of social effect. In 1966 the countercultural impresario Stewart Brand distributed hundreds of buttons asking "Why haven't we seen a photograph of the whole Earth yet?"[165] The poet Archibald MacLeish could not have seen more than the first indistinct television images of the earth taken by the crew of *Apollo 8* when he sat down to compose an essay on how the vantage point of space might remake the self-conception of mankind.[166] "To see the earth as it truly is," he wrote, "small and blue and beautiful in that eternal silence where it floats, is to see ourselves as riders on the earth together, brothers on that bright loveliness in the eternal cold—brothers who know now they are truly brothers."[167] And if the astronauts themselves felt the poignancy of such a view, it was at least in part because they had been primed in advance to take Earth photography seriously. Before their flight aboard *Apollo 8*, Frank Borman, Jim Lovell, and Bill Anders were briefed by Richard Underwood, the NASA official responsible for mission photography, about the logistics of capturing an image of the earth rising over the lunar horizon.[168] Four years later it was Underwood who realized that the nighttime launch of *Apollo 17* would make it possible for its crew, as they traveled toward the moon, to look back and observe a completely round and unshadowed Earth. "That will be the classic picture," he told the lunar-module pilot, Jack Schmitt. "Make sure you get it after you go translunar."[169]

The famous photographs *Earthrise* and *Blue Marble*, then, did not issue from the reflexes of astronauts in epiphany; nor did pictures of the earth taken by satellite or a robot camera onboard the unmanned *Apollo 4* capsule during its flight in November 1967.[170] These images were generated because somebody in NASA predicted they would have an appeal.[171] That prediction proved correct, though it might have been better for NASA had it not. As Robert Poole has documented, the pictures fell from the sky about the same time that astrofuturist assumptions about man's destiny among the stars were yielding to new convictions about the indispensability of the earth. Poole suggests that they did not simply decorate that turn but stimulated it as well: "The sight of the whole Earth gave the world a picture to think with."[172] In particular, in sectors of the ecological movement where the earth, as the ground for all known life, was accorded the status of the sacred, an image of the planet from space could fulfill some of the functions of a religious icon. Probably not many people actually prayed before a poster of *Earthrise*, but their moods might deepen and their minds turn toward questions of ultimate concern, not the least of which was, Who will look after us if we do not look after this? On 22 April 1970 the *NBC Nightly News* presented, without commentary, a montage of the sights and sounds of the first national Earth Day, organized to increase public awareness of environmental issues. The montage concluded by holding for some seconds, in silence, on a photograph of the earth taken from space.[173]

The whole-earth images made available to all a semblance of the perspective offered to the Apollo astronauts as they headed away to the moon; they provided a visual aid and perhaps also a cue to the shared contemplation of man's relation to his planetary home. As Michael Collins observed, however, there was still a significant difference between the viewing experience of the astronaut a hundred thousand miles out and the perception of someone on Earth looking at a mass-produced or telecast snapshot of the same scene. For the astronaut traversing the depths of space, Earth was not ubiquitous, nor was it intentionally exhibited for his gaze: much of the time it was small and difficult to find.[174] And so, insofar as the whole-earth images belong to a category of the sublime, it is the same category that includes late Victorian portraits of the American West. An original apprehension, born of cognitive challenge, rather quickly devolved into an aesthetic appreciation of form and onward into routines of commodification that conveyed a sense of secure possession. "For Earth people only: 12¢ off Tang," ran one magazine advertisement in April 1969. There the planet was, partly shadowed and hanging in space, below it a rectangular coupon to be cut out and used in-store.[175] Images of the earth could work to stimulate a variety of commitments, secular as well as spiritual, and many, like the images themselves, were ultimately paper thin. The planet, it is now clear, has not been saved by its looks.

The earth looked so good in part because the heavens looked so bad. That NASA had been afforded the means to launch its astronauts toward the moon owed much to its success in presenting itself as the modern institutional embodiment of the nation's frontier tradition.[176] But if there was really to be a new frontier in space, the agency's initial exploratory surveys would need to find something that would motivate Americans to pay for its return—a promise of plenitude, like the westward view from the Appalachians, or at least some majestic terrain, a lunar equivalent of the Grand Canyon or Yosemite Falls. As it turned out, there was no plenitude, not on the moon nor on the proximate planets, and the lunar landscapes described by the astronauts and captured on camera during the first Apollo missions seemed arresting only in their monotony; they were safe places for a landing but hardly inspiring to the eye. Later, on the final three missions, the scenery was more impressive—at last, mountains and valleys to match the iconic conformations of the American West—but the audiences were rather smaller by then, and networks were rationing their coverage of moonwalks.[177] The moon had no chance to regenerate earthly culture as, according to popular historiography, the New World had revived the Old at the turn of the sixteenth century and the frontier had remade the republic throughout its first hundred years. Indeed, as some commentators had foreseen, even its traditional poetics were degraded. Sublime and lovely from a distance, the moon now had to endure the indignity of close inspection and derogatory remarks about the quality of its skin. "I'm kind of curious," Jim Lovell informed a press conference after *Apollo 8* returned from its reconnaissance tour, "how all the songwriters can refer to it in such romantic terms."[178]

But perhaps what mattered most was not what Americans saw but the fact that they were watching together. The missions to the moon, just like the early airshows, Lindbergh's passage over the Atlantic, and the flight of *Friendship 7*, could be classed as religious occasions in the Durkheimian sense: a society was attending with moral seriousness to a symbolic representation of itself. Though such occasions were hardly a modern innovation, the advent of national broadcast media did extend the networks of participation and more tightly synchronized their rituals.[179] By the Apollo era, television audiences could watch live feeds from within the spacecraft and from the surface of the moon. And these first expeditions to another celestial body, seeming to augur some radical change in man's relation to his cosmos, had the potential to engage the interest of almost everyone, in the United States and around the world. However, not every minute of every lunar mission proved equally compelling. Over the course of the Apollo program, the country can be said to have watched as one on only three specific occasions, and even in those instances exclusions apply. They were, moreover, *moments* of integration; they did not deepen into seasons of harmony and peace. Nor did these experiences seem so sacramental that the manned exploration of deep space, and the communal witnessing of it, became inscribed into the nation's ritual calendar. By the early 1970s Americans had concluded that these were occasions they could easily do without.

The highlight of the *Apollo 8* mission, for audiences back on Earth, was the crew's prime-time broadcast from lunar orbit on Christmas Eve. While the camera scanned the landscape below, the astronauts offered their personal impressions of the moon and described some of its key features. Then, toward the end of the broadcast, they took turns reading the opening passage from Genesis, concluding with the tenth verse, "And God called the dry land earth; and the gathering together of the waters called he Seas; and God saw that it was good." Frank Borman signed off: "We close with good night, good luck, a Merry Christmas, and God bless all of you—all of you on the good earth." The crew cut the transmission, and abruptly, as if the hand of God himself had touched the switch, television screens across the nation filled with static.[180] At Leonard Bernstein's house a party was in progress; many of those present had little time for the space program. But by the end of the broadcast, the novelist William Styron observed, all cynicism in the room had dissolved, having been replaced by a sense of awe: "It was a moment that was depthless and inexpressible."[181] Two weeks later, the editors of *Christian Century* noted, "We are still aquiver."[182] Richard Nixon evoked the broadcast in the peroration of his inaugural.[183] Yet one commentator in the *New Yorker* detected in himself and in his friends a "secret sense of pettiness and failed imagination." After all the initial excitement, he felt his emotions degrade "from is to was, from delight to gratitude, from absorption to possessiveness, and so on downhill to familiarity."[184] NASA had produced a great show, but the thrill was evanescent, and perhaps, after all, there was no need to see another. Public support for space exploration had increased in anticipation of *Apollo 8*, but it declined again after the crew returned to Earth.[185] In February 1969 only 39 percent of Americans favored a landing on the moon; 49 percent were opposed.[186]

Over the next few months, as the first landing mission approached, anticipation once more warmed public perceptions of the space program.[187] When Armstrong and Aldrin made their descent in the lunar module, negotiating a series of unfamiliar computer alarms and a landing zone clustered with boulders, finally easing to the ground with only twenty seconds of fuel left in the tank, an estimated 123 million Americans, three-fifths of the population, were gathered around their televisions watching the network coverage.[188] After Armstrong's first report from Tranquillity Base, CBS viewers witnessed Walter Cronkite at a loss for words.[189] At home in Orinda, California, Andrew Smith looked at his father and saw tears in his eyes.[190] In the Manned Spacecraft Center movie theater, Norman Mailer, almost despite himself, found that he was happy. It was the same variety of happiness, profound yet slightly disbelieving, that he had experienced years earlier when he had learned that his first child had just been born. Six hours later, as Armstrong stepped down onto the moon, Mailer detected in the theater "a ripple of extraordinary awareness. It was as if the audience felt an unexpected empathy with the sepulchral, as if a man were descending step by step, heartbeat by diminishing heartbeat into the reign of the kingdom of death itself and he was reporting, inch by inch, what his senses disclosed." But as the minutes passed, the mood began to shift. Armstrong and Aldrin, experimenting with their movements, awkward in their exchanges, started to appear a little less extraordinary, and then they had to salute the leaden sentiments of President Nixon. The moonwalk was not yet over, but around Mailer reporters picked up their things and left.[191]

Perhaps the press were more easily bored than the rest of America. But the *New York Times*, seeking to audit the nation's concerns on the same weekend that its astronauts were on the moon, found that many of its citizens remained absorbed in their own parochial cares and routines.[192] In another effort to census opinion, the newspaper elicited reactions from notable thinkers, scientists, artists, and writers, as well as from Jim Aiello, a twelve-year-old boy from Michigan. Like the drinkers in Brewer, not everyone was uplifted. Lewis Mumford declared that popular interest in the mission had been stimulated by "a morbid thrill in the ever-present possibility of a spectacularly violent death." For Reinhold Niebuhr, the landing, though a "tremendous technical achievement," represented "a defective sense of human values." America's impoverished and decaying inner cities, he asserted, merited more urgent attention than the moon.[193] That the flight of *Apollo 11* was a flight from responsibility and therefore hardly an occasion for joy was a common enough sentiment. Black leaders and commentators pressed the point most insistently. Mankind, argued an editorial in *Ebony* magazine, was not yet morally ready to make the rounds of other worlds; it would only become so after it had achieved a "giant leap" in social progress on Earth.[194] Thus, to join the audience for *Apollo 11* was, for many African Americans, to offer some implicit assent to a venture that actually retarded the cause of societal integration. The *New York Times* reported that television sets in bars across Harlem were tuned to a baseball game at the time of the lunar landing.[195]

In broadcasting live the arrival of Armstrong and Aldrin on the moon, NASA engineered a modern equivalent of sacred time: for a few minutes at least, Americans could imagine themselves involved collectively in a cosmic drama. What resulted, in some cases certainly, was an impression of peak experience. But: *post coitum omne animal tristis est* (after sex, every animal is sad). It had been a wonderful occasion but not *that* wonderful. According to a poll conducted in the week following the lunar landing, the majority of Americans had no desire to rush on to Mars.[196] A certain jadedness could be detected even before the triumph was repeated. On the eve of the second landing, in November, a memorandum to White House staff described the likely schedule of events. At 5:52 a.m., it noted, the *Apollo 12* commander, Pete Conrad, would crawl through the hatch of the lunar module and climb down the ladder to the surface: "(You know . . . the same old thing—the Armstrong-Aldrin bit.) (Ho-Hum)."[197] Actually, the second landing would have presented those watching with an altogether different kind of show—Conrad and the lunar-module pilot, Alan Bean, enjoyed each other's company and their time out on the moon much more than the earlier duo had done—but it was canceled early on in the first moonwalk, when Bean accidentally pointed the television camera at the sun and burned out its sensor.[198] The networks switched back to their normal programming.[199]

Would many Americans have paid much attention had *Apollo 13* made it to the moon? None of the networks carried live the first broadcast from the spacecraft, two days into the mission.[200] But only minutes after that broadcast ended, disaster struck, and the voyage abruptly acquired the status of lead item in the news. One oxygen tank had exploded, causing a leak in the second. Without a supply of oxygen, the fuel cells would die and the astronauts too. The only hope was the lunar module, which had an engine that could be used to redirect the spacecraft onto a free-return trajectory around the moon and back to Earth. It also had reserves of air, though these would have to be husbanded carefully. Should the crew survive the perils and deprivations of their looping passage through cislunar space, they would have to trust that the command module's systems were not so damaged or drained of power that it could not carry them safely through the inferno of reentry. It would be more than eighty hours before the television networks could reassure their viewers about the fate of the astronauts, reporting that radio contact had been established and filming the module's gentle descent under parachute toward the South Pacific. In the meantime, the networks—and their audience—maintained an anxious vigil. On 14 April, the day after the explosion, Congress passed a resolution calling on all Americans to pray for the crew, and certainly many did so either in private or at special services.[201] After the astronauts made it home, President Nixon declared a National Day of Prayer and Thanksgiving.[202] For the duration of the crisis, *Time* magazine stated, "a fractured world inured to mass suffering and casual death had found common cause in the struggle to save three lives."[203] But this communion endured no longer than the hosannas that greeted splashdown. Within weeks American troops were at war in Cambodia, and four students were killed at Kent State. And though the recovery of *Apollo 13* may have been NASA's finest hour, the mission overall seemed to drive a deeper wedge between the agency's ambitions in space and those of the nation at large. In May 1970, 48 percent of poll respondents asserted a preference for less government activity in the field of space exploration; only 20 percent wanted more. These levels would remain roughly the same for the next five years.[204]

For ordinary Americans, then, the enterprise of space offered intimations of transcendence but not much more. Certainly the sacred, or at least the sacramental, may have been evoked by the spectacle of launch, by the vision of a distant Earth, and by moments of collective witness as NASA assayed the moon. Yet it was hard to construct a faith out of such experiences when they were shadowed by reservations, widely held, about the merits of the lunar missions. Probably also needed was some kind of sustaining church. NASA was not that, and during the heyday of Apollo there was really nothing else, just small aerospace-industry luncheon groups and the odd short-lived futurist caucus.[205]

Nor was there anything like a literary cult of space, imbuing the voyages to the moon with a philosophy and a poetics that would last a lot longer than the real-time sensations of launch, landing, and return. In the interwar era, many major writers had composed paeans to the airplane. On the whole, that romance had not ended well, and so the next generation, confronted with the rocket, responded with caution, even hostility.[206] These writers felt estranged from the Apollo program for another reason too: they could not journey to the moon themselves. Any effort to capture and convey the primary experience of spaceflight would rely on secondhand information provided by sources they could not fully trust, such as NASA and the news media. There would also be the arduous labor of translation from an original vernacular so tangled in operational jargon, euphemism, and cliché that it could be presumed to have evolved from a fear of actual meaning. For Saul Bellow and John Updike the problem was solved, or at least evaded, by considering the lunar expeditions from the recessed, distracted perspective of protagonists like Artur Sammler and Harry Angstrom, living amid the moral and social chaos of late-sixties urban America. Norman Mailer, in contrast, accepted the challenge in the manner of Henry Adams over a half-century earlier: could he educate himself about the technology of spaceflight to the point that he comprehended the means of its accomplishment without also surrendering to what he viewed as its dead cosmology?

Overall, what the Apollo missions elicited from the leading writers of the age was a poetics of earth and ground, not of machines and starry skies. In a poem written to mark the lunar landing, Archibald MacLeish addressed the moon. "From the first of time," he said, man had thought of it with wonder: "a light beyond our light, our lives—perhaps/a meaning to us . . . ." Then, a reversal: man now stood upon its beaches, beholding the bright Earth: "a light beyond our light, our lives—perhaps/a meaning to us . . ./O, a meaning!"[207] The summer of Apollo over, Rabbit Angstrom is reunited with his wife, and they drive to a motel. The room, "long and secret as a burrow, becomes all interior space," and in a familiar embrace they fall asleep.[208] Mailer, meanwhile, was struggling with his book. He was no longer quite so hopeful that magic and romance could resist the deprecations of technique. "One lifted a book like a boulder out of the mud of the mind," he noted, "and his mind was a pit of wrenched habits and questions which slid like snakes. Where did you put your feet so that finally you might begin?" What he needed was a rock, and eventually he found one back at the MSC, a small gray moon rock, freshly arrived in the world: "There was something familiar as the ages of the bone in the sweet and modest presence of this moon rock, modest as a newborn calf, and so he had his sign, sentimental beyond measure, his poor dull senses had something they could trust."[209] Mailer failed to answer most of his own questions about the space program, but as Laurence Goldstein has observed, he finally "made a virtue of irresolution by affirming that earth is the best place for love, even love of the moon."[210]

Of all the major writers of the period, it was the poet James Dickey who was most confident that space exploration, allied with the right language, could awaken the mind of man to his destiny in the cosmos. Dickey had already described the exultations of flight in his controversial 1964 poem "The Firebombing," which drew on his service as an Army Air Forces radar observer during the war against Japan. The role of the poet, he noted, was "to charge the world with vitality: with the vitality that it already has, if we could rise to it." For its readers, every poem should be "a large, intense and *complete* experience."[211] But after accepting a commission from *Life* magazine to write about the Apollo missions, Dickey seems to have wondered whether his poetry was really very essential to such events. One evening at Cape Kennedy, dining with the *Apollo 7* astronauts a few days prior to their launch, he drank himself into oblivion, as if to mark his own irrelevance before the sublimity of their task.[212] "In a sense they are all poets," he said of the astronauts after their return, "expanders of consciousness beyond its known limits." Indeed, they would accomplish what no earthbound poets could: they plunged "into the universal cavern, into the mathematical abyss, to find us—and return, to tell us what we will be."[213]

However, it did not take too long for Dickey to change his mind. Perhaps, after all, the space age needed drunken poets; and perhaps, after all, it would not remake mankind. Dickey addressed his poem about *Apollo 8* to the lunar astronauts themselves, a gift in verse of their own experiences; by implication, the testimony that they could offer would not be lyrical enough. With the approach of the moon,

>   You lean back from the great light-
> shattered face   the pale blaze
>   Of God-stone coming
>
>   Close   too close

At the end of the poem, rounding the far side, the astronauts travel back into sunlight:

>   And behold
>
>   The blue planet steeped in its dream
>
>   Of reality, its calculated vision shaking with
>   The only love.[214]

The cosmos offered them no invitation—"the void / Is hysterical, bolted out," the moon a "smashed crust / Of uncanny rock"; in contrast, they tremble, enchanted, when they sight the Earth. Dickey's writing is vague here, and not because sublimity always resists definition. He did better in "The Moon Ground," published in *Life* magazine two weeks before the flight of *Apollo 11*. The poem presents the thoughts and feelings of an astronaut out on the lunar surface as he collects samples of moon rocks with his colleague. Their situation is strange—their faces

>   shine
>   Far off, with earth.   We alone
>   Of all men, could take off
>   Our shoes and fly.

But the purpose of their labor reflects the purpose of all human life: to conduct an inquiry into time. As the astronauts bend and pick up stones, they are not just gathering clues to cosmic history, but evoking their own mortality. The human condition has not been transcended by the passage to new worlds. Dickey's astronaut "helplessly" remembers Gray's *Elegy*, set in a country churchyard amid all the stone markers of the dead. There is, he realizes, the same "solemn stillness" here on the moon: "My eyes blind / With unreachable tears." Then he returns to his task: "We bend, we pick up stones."[215]

With the end of the Apollo program and the subsequent confinement of manned space missions to cycles around the earth, the astrofuturist dream was shaded with disappointment, but it did not entirely darken and die. There was, after all, the recent affirmation of big bang theory, with its assertion of an original unity of mind and matter, Earth and the stars. Perhaps the impressions of cosmic harmony experienced across many of the world's religions had been divinations of this deep and latent truth; perhaps the big bang also explained—and could be trusted to sustain—the urge to explore the heavens. As Joseph Campbell observed, "We have actually been born from space, since it was out of primordial space that the galaxy took form, of which our life-giving sun is a member." He judged that "outer space is within us inasmuch as the laws of space are within us; outer and inner space are the same."[216] Space travel, then, expressed a desire for communion. Carl Sagan declared: "We are made of starstuff. . . . The deep human need to seek and understand our connection with the universe is a goal well within our grasp."[217]

But if James Dickey was correct about the power of poetry to charge its readers' world with meaning, then the Americans looking through *Life* magazine on the eve of *Apollo 11*'s departure for the moon would have been in an elegiac mood, mindful of limits more than possibilities, of earthly mortality more than cosmic birth, of stony ground more than open sky. In fact, spiritual responses to spaceflight, in the heavens or back at home, were not determined either by an implicit knowledge of man's kinship with the stars or by the long shadow of the grave. These were universals, when what really seemed to matter, as C. S. Lewis had predicted, was the individual "seeing eye" and the things that were actually seen. Since the orbital flight of *Friendship 7* and John Glenn's matter-of-fact appreciation of a God-given world, the ecology of American religion had grown more vibrant and variegated, and the Apollo program had made accessible an ever-widening range of sights and situations. Amid such diversity there was always scope for an experience in spaceflight to find a meaningful analogue in some cosmology somewhere. But an epiphany applauded in one church might be a heresy in a hundred others and entirely ignored in a thousand more. In the early seventies, no US senator would tell Edgar Mitchell, as Glenn had once been told, that he had inspired a revival of spiritual feelings and values across the whole country. The Apollonian perspective, against some early expectations, had failed to supply a common map to the divine.

That was probably for the best. Pluralism, overall, has been good for American religion. Moreover, though the astronauts did not return from the heavens with the sort of revelations that could dispel religious doubts and inaugurate a new age of faith, they provided substantive proof of the principle that had come increasingly to sustain Americans' adventures in spirit: that experience mattered. No individual epiphany could be claimed as entirely autonomous—Jim Irwin was anticipating a personal encounter with God before he left for the moon—but they aggregated in rather unpredictable ways. The astronauts, after all, had been conditioned to prioritize operational concerns; and if they were to gaze with fascination at anything, it was expected to be the moon, not the earth. But that is not quite how it worked out, and between such expectations and what actually was seen, done, and perceived lay a margin of spiritual and philosophical significance. The capacity of manned spaceflight to recharge the mind with new, unprogrammed thoughts about beauty, order, and scale, about Earth, God, and cosmos, is modest but real. It is one of the better reasons why such missions should continue.

CHAPTER FIVE

Evil Triumphs When Good Men Do Nothing

Religious Americans and NASA in the Autumn of the Space Age

In February 1969 the newly inaugurated President Richard Nixon ordered the establishment of a Space Task Group to develop recommendations for the future direction of the American space program.[1] Six months later, at a meeting in August, Vice President Spiro Agnew, chair of the group, finally came up with an answer to the question that had haunted many of its deliberations: now that man had landed on the moon and achieved the principal objective of the Apollo program, how could the American people be persuaded to maintain their support for space exploration? What was needed, Agnew declared, was a new, "grander overarching goal": "what I propose is a national priority to search the heavens and, before the year 2000, to find God where He lives. We can call the program—'Go for God.'" Some members of the task group were skeptical, with Dr. Lee A. DuBridge, Nixon's science adviser, noting that many theologians considered God to be spirit without body: "We may not be able to find Him, no matter how hard we try." Herbert Klein, White House communications director, was much more enthusiastic: "I think this has PR potential; we can really sell this program. President Nixon, Norman Vincent Peale, and Billy Graham will fall all over each other to back this proposal, and they'll bring every God-fearing American along with them."

Such was the scenario drolly conceived by journalist Bryce Nelson in a short article for the November issue of the *Washingtonian Magazine*.[2] Consistent with comic convention, the article charted the transformation of a plausible situation into something incongruous and absurd. Forwarding the piece to Agnew, NASA Administrator Thomas O. Paine commented, "The

Table 1 Correspondence received by NASA on the subject of religion in space

Year	NASA Figures	Other Reported Figures
1969	185,876	2,500,000—Project Astronaut
		646,000—Gabrielse campaign
		500,000—Fry campaign
		300,000—Texas Jaycees campaign
		35,000—Lynn campaign
1970	901,810	
1971	372,356	
1972	386,169	
1973	1,643,157	
1974		
1975		1,000,000
Total	3,489,368	4,981,000

Space Task Group has been outdone";[3] but he would have recognized that Nelson was tapping, albeit lightheartedly, on a genuinely sensitive nerve. By the late autumn of 1969 the Space Task Group had not succeeded in generating majority popular support for a bold post-Apollo program based on secular scientific and technological objectives. Furthermore, only one constituency of significant size seemed to be actively engaged with the space program at this time, and it was made up of Americans who wished to defend its religious content. Between December 1968 and the summer of 1975, NASA received more than eight million letters and petition signatures supporting the right of American astronauts to free religious expression during their missions in space (see table 1). Of all the topics on which NASA received correspondence in the course of these years, religion in space generated the largest volume of mail, more than four times as much as any other issue.[4] That mobilization, as well as its implications for the standard historical accounts both of American religion in the long 1960s and of the late- and post-Apollo space programs, is the subject of this chapter.

Religious Mobilization in the Space Age

Against the measure of interest and activity offered by the cascade of letters and petitions into NASA offices, all other examples of religious mobilization in the decade and a half between the election of John F. Kennedy and the mid-1970s appear rather modest in scale and force—even the campaigns against the 1962 and 1963 Supreme Court decisions to ban organized prayer and Bible reading from the public schools. Yet the magnitude, the duration, and even the existence of the correspondence campaign in defense of free religious expression in space has generally evaded the attention of historians, even those with a specific interest in rockets or churches. Part of the reason may be that the campaign left few readily visible clues that it had ever occurred. NASA offered little public encouragement to the petitioners, and in most instances the letters and petitions it received were transferred swiftly to a warehouse and thereafter destroyed.[5] In Congress, no Frank Becker or Everett Dirksen emerged, as they had on the issue of school prayer, to give legislative direction and national leadership to the campaign. None of the national councils of the major denominations or evangelical associations adopted the cause as their own, and the most prominent religious journals referred to it at best only in passing.[6] National network television news ignored even the largest of the petitions, and coverage in the preeminent metropolitan newspapers was limited to occasional brief agency reports buried deep on the inside pages.[7] The factors that usually cue historians to take an interest in a subject—the availability of substantial archive materials and a conspicuous public record of contemporary institutional activity, political debate, and media comment—operate unreliably in this particular case.

Even had historians noticed the existence of the campaign, they may not have regarded it as worthy of further investigation. In the 1960s, the secularization thesis encouraged commentators on American religion to immerse themselves in debates about the death of God, the secular city, and the fundamental challenge presented to Christian witness by the problems of race, poverty and war and to dismiss those who sought simply to maintain themselves in their conventional faith, and their country in its, as arrayed on the wrong side of a contest with modernity.[8] Until recently, this same thesis survived, perhaps, to persuade many students of modern America not to choose religious history as their field of expertise, just as related teleological assumptions delayed sustained scholarly inquiries into the enduring phenomenon of twentieth-century conservatism.[9] Although historical monographs on modern American religion certainly continued to be published, they were often produced by scholars animated by a personal religious faith; the diffidence with which the rest of the profession regarded such work was evident in the way that synthetic narratives of the nation's development after 1870 reduced religion to the condition of a "jack-in-the-box," gifting otherwise resolutely secular accounts with intermittent glimpses of the exotic, prayerful other.[10]

In the cases of both religion and conservatism this neglect is now being made good. Yet we still lack a really comprehensive account of Christian grass-roots political activity in the decade following the failure of the Becker Amendment and of its relationship to the conservative movement that emerged with some force—at the time seemingly from nowhere—in the mid-seventies.[11] Hence, perhaps, the tendency for many scholars of the conservative movement to seek explanations for its rise in the realm of secular politics, principally in the backlash against redistributive fiscal policy and affirmative action, and to relegate the significance of religious issues, such as school prayer, to the status of an applause line for George Wallace.[12] Similarly, those scholars interested specifically in the origins of the New Christian Right, though they frequently make reference to a portfolio of grievances assiduously compiled by conservative Christians over the course of the 1960s and early 1970s—the Equal Rights Amendment, access to abortion and pornography, and the prohibition on school prayer—have generally failed to locate much evidence that these grievances were converted into substantial grass-roots religious mobilization prior to the Carter presidency. Indeed, the New Christian Right is often described as the creature of secular conservative strategists who purposefully sought to persuade prominent evangelical preachers, most notably Jerry Falwell, to found and lead religious lobbying groups like the Moral Majority and the Religious Roundtable, thus providing these strategists with a means of drawing evangelical congregations into the broader conservative coalition.[13]

Those who have charted the interplay between conservative political activism and religious faith in the long 1960s, therefore, have tended to regard it as fitful and confined primarily to local or state-level initiatives, such as the 1974 school-textbook dispute in Kanawha County, West Virginia, and the early guerrilla campaigns, in which evangelicals participated, to prevent state legislatures from ratifying the Equal Rights Amendment.[14] Examined in retrospect, after the coalescence of the New Christian Right, these initiatives were usually ascribed the status of a clue: they pointed to the existence of the sort of nascent grass-roots sentiment that could sustain a national movement once it had been created. Yet their very limitations also enforce the conventional wisdom that a national movement had to be created before grass-roots religious mobilization on a national scale could occur. It is that wisdom that meets a challenge in the correspondence campaign in defense of free religious expression in space, for this campaign constituted a nationwide grass-roots mobilization that actually predated the formation of the institutions of the New Christian Right.

This is not to propose, as it were, a reverse trajectory—from spontaneous grass-roots initiative to political institutionalization. There were continuities between the petitions dispatched to NASA between 1968 and 1975 on the subject of religion in space and the subsequent heavy and persistent flow of letters and telegrams received throughout the second half of the 1970s by the Federal Communications Commission (FCC) in defense of the rights of religious broadcasters, a campaign that more directly reflected the strategic interests of leading evangelical organizations and in which their participation can be more readily discerned. Many of those writing to NASA, or later to the FCC, shared with the eventual leaders of the New Christian Right an opposition to the expansion of the liberal, regulatory state and were similarly aggrieved at the failure of that state to protect and promote traditional religious liberties and values. And conceivably, the effect of these two petition campaigns may have been to encourage dreams of a mass movement among those who wished to make politics more responsive to the hopes and fears of evangelical Americans. Both the religion-in-space petitioners and religious broadcasters seemed very pleasurably surprised, even astonished, by the success of their appeals for support.

But the suspicion must remain that the registrations of grass-roots sentiment that occurred in the religion-in-space campaign and even thereafter in the long tide of mail coursing into the FCC were substantially extrinsic to the development of the New Christian Right. After all, most accounts of that movement tend to explain its emergence with respect to other factors and issues, especially the insistence of the Internal Revenue Service in 1978 that private schools that claimed exemption from tax, many of which were run by conservative religious organizations, should prove that they did not discriminate on racial grounds.[15] Moreover, whatever inspiration had been afforded to evangelical political activists by the two mail campaigns found little expression in the work of the New Christian Right once it was actually formed. The New Christian Right was many things, but it was never a truly mass political movement, a point confirmed by the more careful accountings of membership figures for the Moral Majority and its confederate organizations.¹⁶ It could certainly claim some successes, but for all the organizational resources at its command, it never replicated the scale of grass-roots mobilization manifest in the campaigns that preceded its rise.

The religion-in-space petitions offer a corrective to accounts of church-state relations in the long Sixties in which not much of note is held to have happened between the school-prayer decisions and *Roe v. Wade*, and anything that did is presumed to have helped clear the path for Jerry Falwell and his friends. Examined in the light of the conventional birth narratives of modern religious politics, the petitions appear anomalous, their conception underdetermined, their parturition premature. Eight million letters and signatures, however, represent a rather large anomaly to be left unexplained. As this chapter indicates, those organizing the petitions and those who lent their pens to the cause belonged to their times—to the cresting space age in particular—even if they had some doubts that the times belonged to them.

The attitudes of religious Americans toward their nation's space program in the Mercury-Apollo era are, on the whole, more easily inferred than actually demonstrated. If, as most opinion polls in the 1960s suggested, all but a small percentage of Americans classed themselves as religious, then people of faith were well represented among both those who favored maintaining or increasing the budget for space exploration and those who wished to see it reduced.¹⁷ The avowal of religious faith alone evidently did not determine whether a particular individual applauded or deplored the national venture into space. The attitudes of most individuals would have been significantly shaped by secular factors, such as age, gender, race, region, level of education, and political orientation.¹⁸ And to the extent that their views of the space program were influenced by their religious beliefs, the effect of such influence was unpredictable. As we have seen, there was scope for a multiplicity, and indeed a mix, of perspectives. Through the frames of religious faith, spaceflight might appear to be an instrument of human redemption, a means by which a new and better world could be brought into being either on Earth or out in the universe. It might figure, more modestly, as an educational aid that allowed the full glory and cosmic breadth of God's creation to be revealed to man. More modestly still, it might offer periodically the gratification of symbolic success against the atheist Soviet Union in the religious Cold War. Or, alternatively, one could invoke the social gospel and proclaim expenditures on the space program to be basically immoral while there were still poor and hungry people in need of assistance on man's planetary home. The challenges that the space age presented to conventional Christian cosmology may have caused its accomplishments to pass uncheered in some Christian homes. Moreover, NASA was an adjunct of the federal state at a time when that state seemed to be increasingly secular in thought and deed, indeed when it reasonably could be regarded as an agent of the secularizing process in American society at large. There were, therefore, grounds for ambivalence. Americans of faith were not necessarily convinced that the space agency could be counted as a reliable partner in their cause. They were probably right to have such doubts. When NASA officials encountered expressions of religious interest in their activities, even when that interest conceivably could have been converted into a valuable source of political support, they often did not know what to think or how to respond.

The *Apollo 8* Genesis Reading

The ambiguous place of religion within the space program and the ambivalent attitudes of many Christian Americans toward NASA as an adjunct of the secular state are most sharply revealed in the story of *Apollo 8*. This flight is commonly held to have presented the program in its most reverent guise, as its crew, orbiting the moon on Christmas Eve, 1968, concluded a live television broadcast by reading the first ten verses of Genesis. At the time of the mission, all three members of the crew—Frank Borman, Jim Lovell, and Bill Anders—were committed Christians, active in their local churches.[19] Among the items carried on their craft were three Gideon Bibles.[20] While in lunar orbit, Borman also recorded a prayer to be played to his church during its Christmas Eve service, in lieu of the lay-reader duty he had been scheduled to perform.[21]

Of course, the religious contents of the mission were largely incidental to its principal—and secular—purposes. The decision to send a manned spacecraft around the moon was motivated primarily by the need to maintain momentum in the program at a time when problems with the lunar landing module were forcing a postponement of the projected "all-up" test of the complete Apollo system (command, service, and lunar modules) in Earth orbit, a test that had hitherto been regarded as an essential precursor to a lunar-orbital flight.[22] It was not poetics that determined the scheduling of the flight over the Christmas season but rather the readiness state of the hardware and software combined with a launch window that offered optimal lighting conditions for crew observations of the moon.[23]

Nor was it inevitable that the crew's television broadcast on Christmas Eve would contain a religious theme. Although accounts vary in some of their details, it appears that the decision to read from Genesis originated in a telephone call between Borman and Julian Scheer, NASA's assistant administrator for public affairs. Scheer pointed out the likely size of the audience for the broadcast and told Borman that he should "say something appropriate." He offered no further guidance.[24] When the crew discussed the content of the broadcast, Anders, who believed that any message should be "specifically Christian," suggested a reading of the Christmas story, but Borman thought that that was too exclusive, since many non-Christians would also be watching and listening.[25] In the best traditions of NASA, therefore, Borman contracted out the problem, seeking advice from Simon Bourgin, an official from the United States Information Agency who had accompanied Borman on the world tour following his previous mission, *Gemini 7*. Bourgin also struggled with the task, so he called for assistance from his friend Joe Laitin, a spokesman for the Bureau of the Budget. Like Bourgin, Laitin was unhappy with everything he wrote; for example, there was no wish for world peace that would not sound false in light of the Vietnam War. But it was also an issue of style: how to reach for resonance and meaning without succumbing to pretension. Seeking language with "a majestic, almost biblical quality," Laitin, who as a rule was not religious, eventually turned to the Bible, and on the advice of his wife to the start of the Old Testament.[26] There, in the first ten verses of Genesis, he found what he was looking for. Laitin typed up the passage for Bourgin, who subsequently wrote a letter suggesting it to Borman. Bourgin proposed in addition that the reading be preceded by some reflections about the preciousness of the earth when viewed from such a distance.[27] Borman, Lovell, and Anders liked the choice of Genesis, as did Robert Gilruth, director of the Manned Spacecraft Center, and George Mueller, head of the Office of Manned Space Flight, when, "as a matter of politeness," Borman told them about it. "We've been wondering what you might come up with," Gilruth said. "It's a natural idea."[28]

The decision to read from Genesis was indeed natural in the sense that it matched content to the moment—a narrative of earth's creation offered by men viewing it afresh from the shores of another world. It also intimated spiritual significance without enclosing that significance exclusively within the Christian tradition and, perhaps just as important, resolved the problem of style by invoking the authority of the canon. It was also natural because the choice was consistent with the values of the astronauts and senior NASA officials. Yet the choice was not inspired or directed by those values. Borman was at pains to emphasize later that he had never received instructions from anyone in NASA or elsewhere in the government to include a religious statement in the broadcast.[29] For Borman personally, the turn to scripture was less an expression of faith than a pragmatic solution to a basically secular problem: what to say to the world on Christmas Eve. He had Bourgin's letter, with its transcription of the passage, copied onto flame-resistant paper and inserted into the *Apollo 8* flight plan. "I didn't give it anymore thought than that," he recalled.[30]

Certainly Borman and his crew had little opportunity to dwell on the implications of their message during the mission itself. They were busy throughout, performing a constant series of systems checks, preparing their spacecraft for such procedures as translunar injection, lunar orbit insertion, and transearth injection, making manual navigation sightings, photographing the moon and the earth, and studying the lunar surface for evidence of volcanic activity.[31] It was only half an hour before the broadcast that Borman thought to arrange which verses of Genesis would be read by which members of the crew. Still, he stressed to Lovell and Anders the importance of doing it right: "There will be more people listening to this than ever listened to any other single person in history."[32] Moreover, all three astronauts had been genuinely affected by the view of the earth from lunar orbit, an epiphany probably sharpened by their now-intimate acquaintance with what Borman described in the broadcast as "a vast, lonely, forbidding" moon. That "vast loneliness," Lovell declared, "makes you realize just what you have back there on earth. The earth from here is a grand oasis in the big vastness of space."[33] As the time for the reading neared, Lovell switched off his microphone and told his colleagues, "We've got to go into it very nicely." With the television camera trained on the lunar surface as it approached sunrise, Anders informed "all the people back on earth" that "the crew of Apollo 8 has a message that we would like to send to you." He read the first four verses, Lovell the next four, and Borman the final two, signing off with some words also suggested in the letter from Bourgin: "From the crew of Apollo 8, we close with good night, good luck, a merry Christmas, and God bless all of you, all of you on the good earth." Borman turned off his microphone and told Lovell and Anders not to say anything more. The crew then began preparations to leave lunar orbit and head back home.[34]

In the mission-control room at Houston the reading left a number of viewers in tears. "For those moments," recalled Gene Kranz, an Apollo flight director, "I felt the presence of creation and the Creator."[35] The *New York Times* described the reading as "the emotional high point" of the crew's "fantastic odyssey."[36] Editors, noted *Christian Century*, "are supposed to be able to manufacture instant opinions on all subjects, but we are just about struck dumb by this event."[37] No doubt the impact of the reading was amplified by the success of the mission as a whole, but it also contributed to that broader success both by casting the mission in mythic terms (this first manned journey to another heavenly body was not noticeably embarrassed by the implied comparison with the work of God's creation) and by positioning it culturally so that religious Americans could now with assurance claim its achievements as their own. For NASA Administrator Thomas Paine, indeed, *Apollo 8* represented "a triumph of the squares—meaning the guys with crewcuts and slide rules who read the Bible and get things done."[38] That the reading from Genesis had more clearly defined the cultural politics of the space program was affirmed the following month when the *Apollo 8* astronauts addressed a joint session of Congress. After joking that "one significant accomplishment" of their flight had been "to get good Roman Catholic Bill Anders to read the first four verses of the King James version," Borman looked down toward the nine Supreme Court justices, seated at the front, and said: "But now that I see the gentlemen here in the front row, I am not sure we should have read the Bible at all."[39] He recalled later that "the whole chamber rocked with laughter and applause."

Borman's comment, of course, was a reference to the decisions of the Supreme Court in 1962 and 1963 that organized prayer and Bible readings in the public schools contravened the establishment clause of the First Amendment and were therefore unconstitutional. Borman would have been aware that those decisions had been criticized by many of the members of Congress present in his audience (hence the applause) and opposed by a healthy majority of ordinary Americans as well.[40] Although the efforts to pass a constitutional amendment protecting school prayer had largely stalled by the late 1960s, for some the contemporary social history of the country provided incontestable evidence that the court rulings were wrong and a powerful motivation to continue campaigning for their reversal. As one man in Fayette

County, Pennsylvania—where a local school board was defying the court—told a CBS reporter in May 1969, "Look what happened when they took the reading out of the schools, look at the colleges, the riots. The kids today—they don't appreciate nothing. That's because they don't know God."[41] Furthermore, as also intimated by Borman's remarks, there was no elemental legal principle that would necessarily confine the court's recent expansive interpretation of the establishment clause to the gates of the public schools. As Supreme Court Justice William Douglas observed in an opinion concurring with the court's decision in *Engel v. Vitale*, the first of the school-prayer cases, federal and state systems were "honeycombed" with religious exercises funded by the government. He cited the chaplains employed by both houses of Congress and by the US armed forces, the religious services held at federal hospitals and prisons, the US Treasury's inscription of the motto "In God We Trust" on coins and paper notes, and the use of the Bible in the administering of oaths. Such funding, Douglas concluded, "is an unconstitutional undertaking whatever form it takes."[42] Douglas's views were not shared by his fellow justices, but the apprehension that the school-prayer rulings might indeed provide a precedent for further judicial interventions to prize apart religion and the state wherever they embraced did much to animate opposition to the rulings themselves.[43] It was also reflected in the broad protections proposed in the 1964 Becker Amendment, which asserted: "Nothing in this Constitution shall be deemed to prohibit making reference to belief in, reliance upon, or invoking the aid of God or a Supreme Being, in any governmental or public document, proceeding, activity, ceremony, school, institution, or place or upon any coinage, currency or obligation of the United States."[44]

By the time he spoke to Congress, Borman would probably have been aware that the reading of Genesis had indeed excited some critical comment, most notably from one of the successful plaintiffs in the 1963 school-prayer case, Madalyn Murray O'Hair. Since that decision, in which the Supreme Court declared the reading of both the Lord's Prayer and the Bible in the public schools to be unconstitutional—O'Hair had developed what amounted to a proprietary interest in the cause of American atheism, as well as a reputation for intransigence on the question of church-state separation.[45] She argued that the suturing of government authority to Christian belief and practices had the effect of stigmatizing those who were not Christian, whether they belonged to another faith or held no faith at all, as aberrant or simply unworthy of cultural consideration. Thus, as *Apollo 8* made its return to the vicinity of Earth, O'Hair told a Houston radio station that Borman and his colleagues, in reading from the Bible, had been "slandering other religions, slandering those persons who do not accept religion."[46] She announced an intention to start a mail campaign in protest against the reading of prayers and scriptures from space. In some newspaper reports O'Hair was shown sitting in front of a file-card cabinet said to contain the names of twenty-eight thousand "sympathizers." She was holding two file cards in one hand and spectacles in the other, as if poised to dispatch a call for the protest letters at that very moment.[47]

The Response to O'Hair's Complaint

Originating with news agencies and published both in local newspapers around the country and in some of the larger metropolitan journals, these reports of O'Hair's dissent and her putative mail campaign were essentially all it took to provoke a massive and sustained countermobilization in defense of free religious expression in space.[48] Within a day or so of their appearance, letters and telegrams started to arrive at NASA asserting opposition to O'Hair's complaint and appreciation of the Genesis reading.[49] Petitions were circulated within churches.[50] By early January the flow of correspondence into NASA headquarters was such that Thomas Paine ordered that a "thoughtful" set of standard responses be prepared.[51] Local newspapers, meanwhile, had begun to publish details of petitions organized by individuals and groups in their area that expressed disapproval of O'Hair's position.[52] They also published correspondence from readers urging others to write to NASA or to Congress in support of the *Apollo 8* astronauts.[53] In the month of January, James Lynn, of Wichita Falls, Texas, collected an estimated 35,000 signatures for a petition that he submitted to NASA early the following month.[54] Beginning about the same time, Loretta Lee Fry, of Taylor, Michigan, took only two months to gather half a million signatures for her petition, which she presented to NASA officials in Houston on 7 March.[55] By mid-April Mr. and Mrs. Steve Gabrielse, members of a Christian Action Group in Lansing, Illinois, had collected more than 330,000 signatures.[56] Another campaign, Project Astronaut, attracted one million letters and signatures between January and June.[57]

The gathering momentum of the petition campaigns in support of the astronauts' right to say prayers and read from the scriptures cannot be explained by any other actions of O'Hair between her initial remonstrance in

December 1968 and the following summer. There is no evidence that she actually organized a mail drive in protest at the Genesis reading; Frank Borman asserted later that the astronauts had received only thirty-four letters of complaint.[58] She gave a few media interviews, and in late January 1969 she sent a letter to the new attorney general, John Mitchell, asking him to issue an opinion on the constitutionality of the *Apollo 8* mission, given that tax monies had been used to fund "a sectarian prosyletizing [*sic*] activity."[59] It was reasonable to request such an opinion, she asserted, because the alternative, litigation, was so expensive. When the Justice Department demurred, O'Hair did not pursue that alternative, though she promised to do so if similar activities occurred on future space missions.[60] Indeed, O'Hair may have assumed that these warnings had achieved their purpose, for the flights of *Apollo 9* in March 1969 and *Apollo 10* in May were free of conspicuous religious content. O'Hair concluded from this that "there's something going on behind the scenes that we don't know about."[61] She continued to devote the chief portion of her energy and attention to her long-running campaign against tax exemption for churches.[62]

O'Hair only returned purposefully to the issue of religion in space after the *Apollo 11* landing mission in July, for that mission furnished further evidence of NASA's willingness to accommodate Christian liturgies and rituals in its operations. Buzz Aldrin celebrated communion after arriving on the moon and quoted from the eighth psalm in a television broadcast during the journey back to Earth. The same psalm was contained in a parchment given by the Vatican to NASA for photographic miniaturization and inclusion on a disc to be left on the lunar surface. On 5 August, citing these instances alongside the *Apollo 8* Genesis reading and the prayer that Borman had recorded for his church, O'Hair filed a civil suit against Thomas Paine in the US District Court in Austin. In her brief she complained that both in and of themselves and because they involved the expenditure of federal tax revenues, the actions of the *Apollo 8* and *Apollo 11* astronauts, as approved by Paine in his role as NASA administrator, amounted to an establishment of religion. Thus, her constitutional rights as an atheist and as a taxpayer had been infringed. She sought a court order enjoining Paine and NASA from directing or permitting further religious activities in space.[63]

For the first seven months of 1969, therefore, O'Hair herself provided very little additional stimulus to the petitioners seeking to defend the religious expressions of American astronauts. Thereafter, her move toward litigation may have served to clarify the substance of her challenge and indeed to enlarge its significance, for if she had prevailed in the case, the judgment of the courts might well have had implications not just for the astronauts but for all public expressions of faith by government employees in tax-supported contexts. Yet, for the most part, the petition campaigns carried on in a way that indicates a curious obliviousness to what O'Hair was actually doing, in the courts or elsewhere. Correspondence continued to flow into NASA mailrooms after the US District Court dismissed O'Hair's suit in December 1969, after the US Court of Appeals affirmed that judgment in September 1970, and even after the US Supreme Court refused to hear the case in March 1971. Indeed, into the mid-1970s the letters and petitions received by the agency were inspired by the same basically unchanging tale, grounded in a modest early extrapolation of the initial press reports of her protest in December 1968. Whereas initially O'Hair had simply been pictured before a file-card cabinet said to contain the names of twenty-eight thousand sympathizers, now she was said to be actively collecting, or to have already collected, that number of letters and signatures in an effort to coerce NASA into prohibiting any further instances of prayer or Bible reading in space.[64] In 1975 the only novel feature evident in these communications was the spurious attribution to O'Hair of a specific intention to prevent US astronauts on the upcoming Apollo-Soyuz mission from engaging in such activities. "I'm not doing a damn thing," she told a reporter at the time.[65]

What O'Hair did or did not do after December 1968 was therefore of less importance to the petitioners than what she already represented. It is unlikely that there would have been the same scale and intensity of response had any other atheist or secularist protested against the *Apollo 8* Genesis reading. For many American Christians, O'Hair was a figure of near-eschatological significance: they looked at her and apprehended the moral ruin that a secularized world would be. A divorcee who had conceived two children out of wedlock to two different fathers, she had once fled to Hawaii and thereafter to Mexico to escape charges of criminal assault following a confrontation with police over her son's marriage to a minor.[66] She was also fabulously profane. In 1965, during an interview with *Playboy* magazine, she shared these views about Catholic nuns:

> Sick, sick, sick! You think I've got wild ideas about sex? Think of those poor old dried-up women lying there on their solitary pallets yearning for Christ to come

to them in a vision some night and take their maidenheads. By the time they realize he's not coming, it's no longer a maidenhead; it's a poor, sorry tent that *nobody* would be able to pierce—even Jesus with his wooden staff.[67]

O'Hair was similarly unsparing in her expressions of disdain for the Christian faith more generally and for those who held to it. In a letter published in *Life* magazine in 1963 she described the Bible as "nauseating, historically inaccurate, replete with the ravings of madmen. We find God to be sadistic, brutal, and a representation of hatred, vengeance. We find the Lord's Prayer to be that muttered by worms grovelling for meager existence in a traumatic, paranoid world."[68]

All of this could have been discounted as nihilistic provocation had O'Hair never once achieved her purposes. Yet, on the issue of prayer and scripture reading in the public schools her purposes had indeed been achieved. That the Supreme Court had in the past lent its authority to her cause deepened the anxieties of her critics as they tried to predict the outcome of her subsequent campaigns. Nevertheless, none of her previous endeavors, not even her case against school prayer, had conjured up such a storm as her protest against the reading of Genesis in space. Without really realizing it, O'Hair had executed her most perfect provocation. Here was an agency of the federal state almost archetypically identified with secular scientific and technological concerns engaged in an epochal act of space exploration and marking the occasion with an attestation of God's creative power at a time when many Americans feared that religious beliefs and values were losing their force. Then, to instantly dissolve whatever satisfactions and comforts the Genesis reading had allowed, here came the apostle of a godless world announcing her intention to prevent such a consecration from ever happening again. If O'Hair was permitted to achieve this prohibition, what would be left in public life that could be held securely sacred? It appeared that a critical juncture had been reached.

Hence the running start achieved by the petition campaigns, and hence also their scale and longevity. In 1969 four large petitions expressing support for the right of US astronauts to speak freely of God were presented to NASA (see table 1). The largest, submitted in September and by then comprising 2.5 million letters and signatures, was generated by the Project Astronaut campaign, organized by the fundamentalist Family Radio Network in California.[69] Next in size were the petitions of the Gabrielses, who submitted

646,000 signatures in December, and Loretta Lee Fry, who submitted 500,000 in March.[70] In October, the Texas Jaycees announced that they had collected more than 300,000 signatures.[71] Not registered in these larger campaigns were the letters and telegrams sent to NASA by individual Americans and independent petitions circulated among church congregations and local civic groups.[72] In addition, NASA was not the only recipient of mail on the issue: opponents of O'Hair also dispatched letters in liberal measure to the White House, to their congressional representatives, to their local newspapers, and even to the US attorney in Texas who was responsible for preparing Thomas Paine's defense.[73] Moreover, the flow of correspondence continued into the mid-1970s, with notable peaks in 1973 and early 1975. In 1973, according to NASA Administrator James C. Fletcher, the Manned Spacecraft Center in Houston was "sorely taxed" in its efforts to deal with the volume of mail it was receiving on the subject.[74]

In a letter to Thomas Paine written on the day that Paine was to be presented with the Project Astronaut petitions, Harold Camping, president of Family Radio, suggested that the campaign in support of the *Apollo 8* astronauts represented "the largest voluntary commendation of an act by man that has ever occurred in our nation's history and perhaps in the history of the world."[75] When the Project Astronaut petitions are counted alongside the evidence from the other correspondence campaigns of 1969 and beyond, yielding a total of at least eight million signatures and letters between January 1969 and July 1975, that statement hardly seems hyperbolic.[76] Most strikingly, the measures of correspondence on the issue of prayer and scripture reading in space exceed by some degree the most commonly cited figures for the mail campaigns that followed the court decisions to ban such practices within the public schools. After *Engel v. Vitale*, the Supreme Court received five thousand letters, most of them critical, but it received fewer than one hundred after *Abingdon School Dist. v. Schempp*. About a million signatures were received by the House Judiciary Committee in support of a school-prayer amendment.[77]

The apparent difference in the scale of popular mobilization in support of prayers in space compared with prayers in school may reflect the divergent positions adopted on the two issues by many major national institutions and organizations. In the case of school prayer, opponents of the prohibition were forced to contend not just with the judgment of the highest court in the land but also with the endorsement of that judgment by a number of religious authorities, including the National Council of Churches, and the assertion of the president that it should be respectfully accepted.[78] In contrast, O'Hair's attempt to constrain the religious expressions of US astronauts was practically an orphan. Even the organizations that were usually most assiduous in their efforts to construct and maintain the "wall of separation" between church and state—the American Jewish Congress, the American Civil Liberties Union, and Americans United for Separation of Church and State—elected to let O'Hair work alone on the space-prayer issue. Although local ACLU members did consult with O'Hair, and although the union's national legal director believed that her protest stood "on sound theoretical ground," it lent no public support.[79] Americans United, meanwhile, even declared an interest in filing an amicus curiae brief on behalf of the defense in *O'Hair v. Paine*, taking the view, as did the American Jewish Congress, that the astronauts had been exercising their constitutional right to freedom of religion.[80]

Still, if O'Hair lacked political and institutional allies in her protests against religion in space, so did her opponents, making the reach of their petitions even more impressive. The White House kept its distance.[81] The major metropolitan newspapers on occasion criticized O'Hair—"Madalyn Needs Padalyn," commented the *Detroit Free Press*—but the correspondence campaigns in support of the astronauts were reported on only rarely, and then, with one notable exception, as a curiosity, not as a cause.[82] Local congressmen wrote polite responses to concerned constituents, entered letters and local newspaper editorials into the *Congressional Record*, fielded the requests of petitioners for appointments with NASA officials, and in some cases attended their presentations, but none sought to adopt the issue as their own in the manner that Frank Becker had with respect to school prayer.[83] Only in the case of the Jaycees in Texas and later Ohio and the American Legion in California in 1974 could those campaigning in support of religious expression in space draw on the resources of established organizations.[84]

For the most part, the petitioners seem to have begun with rather limited ambitions, working through networks in their home communities, using local print and broadcast media where possible, in the hope of collecting a few hundred or a few thousand signatures. James Lynn, for example, started circulating a petition among his friends, neighbors, and coworkers in Wichita Falls, gathering 150 signatures—"an amazing number," he said—within a few days. After a report about his efforts was published in a local newspaper, his ambitions scaled up to a total of 5,000. The report, which gave his home address, prompted others to circulate the petition and send him the signatures they collected. He was assisted, in addition, by a column that appeared in a number of Texas newspapers; it printed a copy of the petition to be cut out, signed, and mailed to the columnist for forwarding to Lynn. In this way the campaign quickly spanned out from Wichita Falls to include other Texas towns, as well as communities to the north in Oklahoma; petitions arrived also from Louisiana, Idaho, Arkansas, and California. Often the petitions had been circulated among local church congregations, civic groups, and workplaces: employees of the Scurlock Oil Company in Houston, for example, contributed 40 names to the cause. By early February, Lynn had collected about 35,000 signatures to send on to NASA.[85]

Loretta Lee Fry, meanwhile, was the daughter of a Pentecostal minister and a mother of four who broadcast a daily Bible question-and-answer radio program from a station located in the window of a Christian-goods store in Southgate, Michigan. Learning of O'Hair's threatened mail campaign against the reading of Genesis, Fry appealed to her audience—perhaps recently enlarged as a result of a profile published in the *Detroit Free Press*—to send in petitions expressing support for the astronauts. "The response has left me breathless," she told the *Detroit News*, which itself had aided her campaign with an early sympathetic report. She printed out enough petitions for 40,000 signatures and then 60,000 more. Fry's petition drive was joined by other religious broadcasters in northern Michigan and Ohio. By the end of February, Fry had amassed more than 400,000 signatures; by the time she presented the petitions to NASA on 7 March the total exceeded half a million.[86]

Of the four large petitions of 1969, it was probably Project Astronaut that began with the grandest aspirations—a goal of 100,000 letters—as well as the widest geographical reach. The Family Radio Network owned stations on both coasts, in California and New Jersey, and its programming was also broadcast on other stations around the country.[87] Once again, however, expectations were emphatically exceeded, as Project Astronaut attracted support not just from beyond the range of Family Radio's transmitters but from outside its primary fundamentalist audience. Information about the campaign, as well as exhortations to join it, passed from Family Radio listeners to successive relays of local and denominational newspapers, often receiving editorial endorsements along the way, and also into grass-roots social and associational networks.[88] In July 1969, for example, members of the Sunset Hacienda Woman's Club in San Gabriel Valley, California, voted to circulate the Project Astronaut petitions themselves and to encourage other woman's clubs in the valley to do the same.[89] In Stamford, Connecticut, the travel agency run by Andrew Robustelli, a former New York Giants defensive end, advertised Family Radio's campaign in the *Stamford Advocate*.[90] Such local initiatives continued even though Family Radio itself had ended its appeal in May, having met its revised target of one million letters and signatures. By September, when these letters and signatures were presented to Thomas Paine, they totaled more than two and a half million and were still arriving at Family Radio's offices in San Francisco at a rate of about twenty mailbags' worth a day. "We believe that Family Radio's leadership in this effort has been negligible," Harold Camping told Paine. "Rather this is almost completely a spontaneous expression from the people of our land."[91]

The petition campaigns in defense of the *Apollo 8* Genesis reading were spontaneous not in the sense that they sprang from nowhere but in the sense that they were neither a creation of nor controlled by any single organizational entity or formal alliance of entities. Rather, they welled up in many places from under common ground. There is evidence of participation by Catholics as well as Protestants, by evangelicals and nonevangelicals, by Baptists, Presbyterians, Lutherans, and Methodists, and by civic groups—most obviously the Jaycees and the American Legion—with no specific denominational allegiance.[92] Certainly, fundamentalist Christians, the principal audience for Family Radio, may have felt a particular inclination to defend the reading of Genesis, the literal truth of which was a point of contest both in emerging state-level skirmishes over the content of school biology textbooks and in ongoing debates within the conservative evangelical community about the necessity of an inerrant Bible.[93] But though a faith in the very letter of the scriptures was manifest in some contributions to the campaign—"I am a Christian and believe every word in the Bible," one correspondent told Paine—it was also entirely possible for participants to discount the veracity of the Genesis story and still cherish its reading as a ritual expression of the belief that however he had done it, and to whatever time scale, God had created the heavens and the earth for man.[94] It was apparently in this spirit that the *Apollo 8* astronauts themselves viewed their chosen text.[95]

For the most part, indeed, the letters and petitions made their stand not on a specific point of scripture but on an article of the constitution, the First Amendment protections for the free exercise of religion and the freedom of speech. "It is our opinion," asserted the signatories of one early petition, "that Astronauts Anders, Borman, and Lovell were within their constitutional rights and the limits of good taste in reading the Genesis account of creation, as well as expressing themselves in prayer."⁹⁶ However, for many of those who contributed their names to the campaign, it was not just the rights of astronauts that might wither under challenge from Madalyn Murray O'Hair. At stake was the right of Americans to hear the word of God, from space or in any other public setting. "I think <u>we have every right to hear God's name in public</u>," wrote one correspondent, "and I do not believe it should be hushed up because of one woman."⁹⁷ A petition received by the White House in March 1970 asked, "Why should one or a very small minority of Americans (?) (?) be allowed to impose their will upon the Vast Majority of true Americans, who also have the <u>right of free speech, freedom of religion</u>, etc."⁹⁸

In the marrow of such apprehensions—what compelled these assertions to be underlined—lay a memory of the victory that O'Hair had already achieved on the issue of school prayer. In a letter to his local newspaper, later inserted in the *Congressional Record*, George Levesque, of Elmira, New York, declared that if O'Hair were to succeed with respect to religious speech in space, she would probably seek thereafter to remove the phrase *under God* from the Pledge of Allegiance, to prohibit oaths on the Bible, all benedictions and invocations at public events, and use of the US Postal Service for religious mail, and also to eliminate Christmas as an official national holiday. "This list may sound ridiculous," he wrote, "but so was the prohibiting of school children from repeating a prayer in public school, and so is what Mrs O'Hair is trying to do now."⁹⁹ Another letter writer noted: "Mrs O'Hair had much to do with taking prayer out of public schools. Are we, as Christians, going to sit back and let her take the word of God (Our Creator) from everything?"¹⁰⁰

What many American Christians now expressed with respect to the school-prayer decision, indeed, was a memory of their own passivity. The president of the Texas Jaycees declared, "We did nothing as an organization to stop her from removing religion from our public schools and she was successful."¹⁰¹ Here, then, was one certain lesson to be learned: "Evil triumphs when good men do nothing," read the legend emblazoned across the petitions submitted to NASA by Loretta Lee Fry.¹⁰² It was a lesson that contained a breath of heresy, for many conservative Christians had hitherto believed that in preparation for the imminent Second Coming of Christ, good men should be attending, not to the affairs of politics and government, but to their own relations with God. Yet, if they permitted O'Hair to achieve what were imagined to be her purposes, if they allowed the Word to be cast out from the discourse of the state, could they still expect to be summoned unto heaven? Could they still even call themselves "good"?

Certainly, the institutions of government, left to their own devices, could not be trusted to do the right thing. That O'Hair had been successful in her pursuit of a ban on school prayer was now cited as evidence of an impoverishment of spiritual conviction and moral will at the center of national politics. "I wonder why we don't have statesmen and politicians that will stand up and speak for God?" asked a letter published in a West Virginia newspaper. "If we don't complain who will?"[103] Some, indeed, assumed that prior to their departure the crew of *Apollo 8* had not "whispered a word" to anyone of their intentions with respect to the Genesis reading. If they had done so, the thinking went, a fear of constitutional entanglement would have caused government officials to veto the plan.[104] For others, the state, and NASA specifically, was not beyond promise of redemption. The Genesis reading, after all, had been quite a gift to Americans of faith, and what would have been the point of the correspondence campaigns if there was no prospect that someone in authority was prepared to be persuaded? Yet these institutions were regarded at best as unreliable allies, potentially willing to surrender the farm to secular humanism unless they were fortified by clear evidence that public religion had public support, and, as a necessary concomitant, implicitly cautioned that any such surrender would carry real political costs.[105] As Harold Camping told Paine, "These letters are telling us that a major segment of our citizenry is deeply interested in the space program but that it can only be successful if it has God's approval and if it is kept in a right relationship to God." He went on: "May we encourage NASA to continue to acknowledge God and to give Him the honor and the glory as they proceed with new exploits."[106]

The Afterlife of the Petitions

The petitions were received politely by NASA officials and then hastened quickly into warehouses. Those who had organized the campaigns, meanwhile, appear to have been viewed as well-meaning naifs, and not—as perhaps they should have been—as conduits to a potentially valuable reservoir of public support. In 1969 religious observances remained at best a contingent feature of NASA's activities, incidental to its immediate purposes. As one public-affairs officer commented blithely when presented with the petitions collected by Loretta Lee Fry, "We're always looking for spinoff from the space program, and if religion is one of them, that's great."[107] Although the lunar landing mission of *Apollo 11* was marked by conspicuous acts of public religiosity, these acts generally occurred at the initiative of President Nixon, as advised by Frank Borman, who was then serving as a NASA liaison in the White House; they were not the product of the agency's own public-affairs machine. It was Nixon who asked Borman to reprise the reading of Genesis at a White House worship service as *Apollo 11* neared the moon; he also issued a proclamation calling on all Americans to pray for the successful and safe conclusion of the mission; and, following a recommendation from Borman, he greeted the astronauts on board their recovery ship, the USS *Hornet*, by inviting the ship's chaplain to say a prayer of thanksgiving.[108] According to Julian Scheer, Nixon ordered that the phrase *under God* be included on the plaque to be left on the moon, as in "We Came in Peace Under God for All Mankind," but the instruction was ignored. Scheer assumed that "in the rush of events, no one would remember."[109] Buzz Aldrin's request to carry communion elements—wine, a chalice, and a wafer of bread—to the moon was approved by NASA, but he was asked by Deke Slayton not to read the communion passage, from the book of John, back to Earth. This condition Aldrin attributed to the agency's nervousness about Madalyn Murray O'Hair.[110]

Aldrin observed radio silence during his celebration of communion, but O'Hair still decided to take NASA to court, which probably chilled still further the agency's enthusiasm for entertaining religious practice and speech on its spacecraft and other facilities. NASA's legal representatives were optimistic that they would prevail in the case, but they also noted that between the establishment clause of the First Amendment and its protection of the free exercise of religion existed "a gray zone of uncertainty" about the kinds of religious activities that were permissible under the auspices of a federal agency.[111] For the same reason, and because it was anxious to avoid a second lawsuit, NASA declined to grant an easement of land at the Kennedy Space Center to a private corporation that wished to construct a "Chapel of the Astronauts" as "a national symbol of the link between man's penetration of outer space and his faith in God."[112] The courts eventually dismissed O'Hair's complaint on grounds that should have given the agency confidence that any similar suits in the future were also unlikely to succeed: that taxpayer status did not give a plaintiff sufficient standing to contest the constitutionality of low-cost, non-congressionally mandated religious activities occurring within a primarily secular government program.[113] Nevertheless, NASA refrained from exploiting the latitudes that could be inferred from the judgment for the purpose of drawing Christian caucuses more securely into its political base. Although astronauts continued to take religious artifacts into space (*Apollo 14*, for example, carried a hundred microfilmed copies of the Bible to the surface of the moon), there was nothing liturgically equivalent to the Genesis reading or Aldrin's lunar communion in the rest of the Apollo program.[114] During *Apollo 15*'s sojourn at the foot of the Apennine Mountains, Jim Irwin suggested to mission commander Dave Scott that they conduct a religious service, but Scott demurred.[115] NASA managers might well have been thankful for that. Mounting a legal defense in the case of *O'Hair v. Paine*, though it turned out successfully, had still been a distraction and a bore. Busy astronauts and officials had to find time to be interviewed by counsel, search their files for evidence, and submit affidavits.[116] The agency probably was not eager to invite a repetition of the ordeal.

Thus, even though NASA, having essentially dissolved the principal justification for its own existence by winning the race to the moon, was particularly in need of public displays of affection and support, it elected not to foster the constituency of Christians writing to it in their millions. The agency instead endeavored to defend the Apollo program against budget hawks within the Nixon White House by emphasizing the scientific discoveries to be made on the moon and to generate popular enthusiasm for an eventual manned mission to Mars on the grounds that it would yield new knowledge and technological innovations that could be applied to improve the conditions of life on Earth.[117] Moon rocks, however, had a value that was difficult to subject to persuasive cost-benefit analysis, and perhaps inevitably therefore, it was the costs that became most persuasive. Many Americans, meanwhile, regarded the discharge of resources into deep space as an eccentric way to make advances back at home.[118]

In the mid-1970s, with correspondence defending religion in space continuing to course into its mailrooms and with Jim Irwin embarking on his own High Flight evangelical ministry, NASA officials seem finally to have registered both the existence of a Christian audience for its activities and the political rewards that might accrue from a public-affairs program tailored to this audience's specific interests and concerns. NASA Administrator James Fletcher, himself a devout Mormon, believed that "the churches are our strongest supporters" and that mobilizing their goodwill could have "a big public impact."[119] John Donnelly, NASA's assistant administrator for public affairs, also noted the importance of reaching "the religious market." Preachers, he asserted, "are a key multiplier factor, but we've never really organized a program to get the message to them." He proposed that the agency consult the religion editors of major newspapers and journals in order "to determine the best channels for reaching clerics, getting their views on the views of preachers toward the program, what *'messages'* would most appeal to them and how we can spread the good over effectively among this audience."[120] Evidence that any such program was actually initiated has so far proved elusive. Even if it was, it happened too late, for the season of optimal effect had passed years earlier, in 1969 and 1970. It was then that the essential parameters of NASA's activities and budgets for the next decade had been negotiated and were thus most pliant to public intercession; and it was then that millions of religious Americans had offered their support to the agency, only to be viewed with polite indifference.

Meanwhile, those who had organized the petition campaigns in favor of religion in space largely failed to parlay their successes on this specific issue into any broader, more sustained Christian political movement. Project Astronaut evolved into Project America, with the objective of returning prayer and Bible reading to the public schools, but then swiftly disappeared from view.[121] Still, there were legacies of a kind. Having sought a prohibition on the reading of scripture in God's very heavens, O'Hair could now be accused of intending virtually anything, and to millions of Americans the charge would seem plausible. In mid-1975, just as the flow of correspondence on the subject of religion in space started to recede (following Apollo-Soyuz, there were no manned US space missions for another six years), rumors began to circulate that O'Hair had petitioned the Federal Communications Commission for a ban on all religious radio and television programming. The rumors were false; although a petition had been submitted to the FCC with respect to religious broadcasting, O'Hair was not its author, and its objectives were rather modest. In December 1974 Lorenzo Milam and Jeremy Lansman, two community-radio activists based in California, asked the FCC to investigate whether religious institutions holding licenses to broadcast on restricted educational FM and television channels were adhering to the fairness doctrine and also whether the programming on these channels was genuinely educational. In addition, they requested that for the duration of the investigation the commission establish a freeze on new licenses of this kind.[122]

The petition was denied in August 1975, but as with the dismissal of

O'Hair's suit against Thomas Paine, this procedural closure went unnoticed or unheeded. Before the decision, about 700,000 letters had been sent to the FCC calling on it to reject the alleged petition from O'Hair; by November the total had reached 1.5 million, and by February 1976, 3 million.[123] The imperviousness of these protests to what had really occurred—again, like that of the correspondence campaigns in defense of religious expression in space—can be attributed in part to the way they spread, in large measure along grass-roots networks. The rumor concerning the petition, along with entreaties to write to the FCC, traveled from local church to local church, local newspaper to local newspaper, across points on the map not policed by knowledgeable and attentive authority.[124] But the most compelling evidence of a continuum between the two campaigns lies in the actual content of the mail received by the commission. Many of those writing asserted that O'Hair had submitted 27,000 letters or signatures in support of her petition, the same figure that had been cited in much of the correspondence sent to NASA and a figure itself drawn from early reports of her original complaint against the reading of Genesis by the crew of *Apollo 8*.[125] Others were under the impression that the point of O'Hair's submission to the FCC in fact had been to advance an objection to that reading, more than seven years after it took place.[126] The confusion was further compounded in at least one flier circulating at the time, which alleged that O'Hair had issued a demand to NASA insisting that the agency, presumably by means of a dramatic arrogation of new powers, prohibit any prayer or mention of God on television.[127]

The FCC correspondence campaign was not, however, solely the outcome of spontaneous local initiatives spreading outward along parochial religious and associational networks. In contrast to the defense of scripture reading by American astronauts, the defense of the prerogatives of religious broadcasters reflected the immediate concrete interests of many religious organizations in the era of the "electric church." Some of these organizations had filed formal responses to the Milam-Lansman petition with the FCC.[128] It was a commonplace assumption within evangelical circles in the 1970s that television and radio offered a revolutionary new means of preaching the gospel and of generating a revival of religious faith and moral values; conversely, it was easy to imagine the eschatological consequences of churches' being denied the right to broadcast and control of the airwaves falling into the wrong hands.[129] The leading evangelical organizations in particular were determined to resist any attempt to restrict or in other ways interfere with their broadcasting ventures. National Religious Broadcasters, at that time the broadcasting arm of the National Association of Evangelicals, sought to inspire and coordinate vigorous opposition to the Milam-Lansman petition among its membership, representing it as an existential threat to all religious stations, educational or not.[130] And under the front-page headline "Yes . . . She's At It Again!" the *Christian Crusade Weekly*, the newspaper of Billy James Hargis's fundamentalist Christian Crusade, made what seems to have been the first connection between O'Hair and the petition. The paper informed its readership that O'Hair had "taken the lead in the drive that could eventually see the end to all broadcasting of religion on radio and television," including Hargis's own radio ministry.[131] (Later, the *Weekly* attributed the rumor of O'Hair's involvement to National Religious Broadcasters' June 1975 newsletter, where no actual reference to O'Hair appears.)[132] It announced a campaign to obtain one million signatures in opposition to the petition, printing counterpetitions for the purpose.[133] Most likely these were the springs from which the prodigious cascade of correspondence began to flow. Through the 1980s the letters continued to arrive at the commission at a rate of about 15,000 a week; by 1986 it had received 25 million items of mail on the issue.[134]

In June 1972, recently resigned from the NASA astronaut corps and preparing to embark on a full-time religious ministry, Jim Irwin told a news conference at the annual Southern Baptist Convention of his disappointment that Madalyn Murray O'Hair had raised no objection when he had spoken a line from Psalm 121 while walking on the moon. Her earlier opposition to the reading from Genesis, he asserted, "had generated much beneficial public support" for the space program.[135] Of course, it was not quite as simple as that. What the petitions in defense of religious expression in space disclosed, as did the later correspondence received by the FCC, was the complex mix of goodwill and apprehension with which religious Americans regarded modern technology under the charge of the state. When astronauts in lunar orbit gave a reading from the scriptures, and when preachers on an FM frequency issued a call for prayer, the work of such technology could appear to be providential, for it amplified the Word and promised multiplications of the kingdom of the faithful. But church-state jurisprudence being what it was, these were blessings that could conceivably turn overnight into their own negation. If the courts so decided, both the airwaves reaching to each earthly horizon and the infinite expanse of the heavens above might be cleared of Christian speech. Even without a formal court decision, federal agencies perhaps would take the initiative themselves, with much the same profoundly secularizing effect. In the forebodings of religious Americans, O'Hair was certainly a conspicuous cause for concern, but she could only achieve success if the institutions of state lacked the conviction to resist. The petitions flowed from the anxiety that NASA's heart was not in the fight. Though it had never been "Go for God," the space program had periodically seemed—for example, immediately following the flight of *Apollo 8*—worth an investment of Christian hope. But now there was nothing in NASA's attitude that gave assurance of such investments' ever receiving a return. In the autumn of the space age, NASA appeared to symbolize the future of the federal state writ large, complicit in the desacralization of the public sphere and enclosing the very heavens in the dominion of the profane.

Epilogue

Twenty-four times the shuttles had launched. Twenty-four times they had returned safely to Earth. By the twenty-fifth shuttle mission successful launches had come to seem routine, to the American public at least. Thus, on 28 January 1986, when the space shuttle *Challenger* rose from an ice-encrusted Cape Canaveral into the Florida sky, the commercial television networks were busy with other things.[1] But there were people watching, mostly on CNN or NASA TV, and many of them were children sitting in their schools, for in addition to its professional astronaut crew, *Challenger* was carrying Christa McAuliffe, a teacher from Concord, New Hampshire. The previous year, McAuliffe had been selected as the winner of NASA's Teacher In Space Program, and in the course of the mission she was scheduled to conduct educational experiments and also to deliver two lessons on the subject of spaceflight for broadcast live to the nation's public schools.[2] McAuliffe was the first private US citizen to travel on a NASA mission. It seemed likely that others would follow her soon. The agency had already announced its intention to carry a journalist on a future shuttle flight.[3]

In the office of Peggy Noonan, a White House speechwriter, the television was on and tuned to CNN. But Noonan, who was in the middle of a phone conversation, was not watching when, seventy-three seconds after lift-off, the screen showed the rising shuttle abruptly bloom and disintegrate. An assistant rushed in with the news. Viewing the unfolding coverage, Noonan quickly realized that the crew were lost and that President Reagan would have to talk to the nation. She began drafting his remarks, which would draw on notes from a conversation between Reagan and a group of network news anchors an hour or so after the tragedy occurred. There were a few other significant contributions and revisions before Reagan delivered the speech from the Oval Office at five o'clock that evening, but not nearly as many as for most presidential addresses. The closing passage of the speech, with only minor editing changes, survived intact from Noonan's original draft: "The crew of the space shuttle *Challenger* honored us by the manner in which they lived their lives. We will never forget them, nor the last time we saw them, this morning, as they prepared for the journey and waved good-bye and 'slipped the surly bonds of earth' to 'touch the face of God.'"[4]

Noonan does not explain how she came to include the quotations from "High Flight" in the address; she only recalls her intuition at the time that Reagan would know the poem: "It was precisely the kind of poem he would have known, from the days when everyone knew poems and poets were famous, everyone knew Robert Frost and Carl Sandburg. It had been popular during the war." She was right. Reagan told her later that the poem had been inscribed on a plaque outside his daughter's school. She also learned from a Hollywood press agent that Laurence Olivier had read "High Flight" at the funeral of Tyrone Power, who, like John Gillespie Magee, had served as a pilot during the Second World War.[5]

Perhaps it had always been thus: the melancholic knowledge of Magee's own early death and a growing association with mournful experience as a result of its recitation at the funerals of aviators combined to convert a poem that abounded with the joys of life into a kind of requiem, with its exultant closing line now read as a synonym for death. Still, until the *Challenger* disaster the poem seems to have retained its original optimistic meaning for the pioneers of spaceflight, expressing a hope of transcendence implicit in the experience of spaceflight itself, not a consolation for times when a mission went fatefully wrong. When Pat Collins typed Magee's verse onto a small card for inclusion in her husband's personal-preference kit and when Jim Irwin appropriated "High Flight" as the name of his ministry, they had no intention of evoking the grave.[6] "We're still pioneers," Reagan assured the nation. "We'll continue our quest in space." But by the end of his address, even before inquiries into the disaster revealed serious deficiencies in the design of the shuttle and in the decision-making process that allowed *Challenger*'s launch that icy morning, something had changed.[7] Into the mid-1980s Magee's poem remained the best expression of the thought that spaceflight could be a benediction. After Peggy Noonan had done her work, the poem had a different meaning: spaceflight as both metaphor and occasion for the migration of the soul after dying. Spaceflight had brought the crew of *Challenger* closer to God, but only by killing them first.

The *Challenger* mission, with McAuliffe on board, had been conceived as a means of reversing the attrition of public interest in the space program since the heyday of Apollo. The shuttle, unlike earlier spacecraft, had room for "citizen passengers," who could focus on the task of communicating the experience of spaceflight to ordinary Americans without the distraction of operational requirements, in language that presumably would be free from the technical jargon that punctuated the speech of professional astronauts. After the teacher, there would be a journalist, and then who knows? John Denver and George Lucas had both expressed an interest in traveling on the shuttle.[8] Perhaps Michael Collins would finally see his triad of philosopher, priest, and poet aloft in the heavens.[9] Would the courts have permitted a minister aboard? Would they have licensed a planned, public performance of a religious service in space? If so, and even if not, there would have been interesting effects. When Christa McAuliffe took her seat on the mid-deck of the shuttle, she was opening up a future in which spaceflight could conceivably become relevant once again to Americans of faith *because* of their faith.

But it was not to be. After the disaster, the shuttle fleet was grounded for almost three years, and the application and selection program for citizen spaceflight participants was suspended.[10] No citizen passenger would travel aboard an American spacecraft until John Glenn returned to orbit on the shuttle *Discovery* in October 1998, and he was obviously a rather special case. Thus, through the post-Apollo period, religion and the space program met mostly in rites of mourning, as after *Challenger* and again, in 2003, when its sister ship *Columbia* was lost on reentry.[11] "God bless and bless and bless their souls," Peggy Noonan wrote of the dead *Columbia* astronauts, "and rest their souls in the morning."[12]

A generation earlier, spaceflight and spirituality had combined in a relation rich with both promise and portent. Most accounts of the US space program in the Mercury-Apollo era address little more than a desultory paragraph to the theme of religion. A few, in conscious contrast, thrill to the provocation of pronouncing the program religious in its aims and dominant values. In truth, spaceflight was a field of human endeavor in which conceptions of the sacred pressed up against apprehensions of their own negation.

From the teasing assertions of Soviet cosmonauts that there was no evidence of God in space, through the reading of Genesis during the first manned American mission to the moon, to the final wave of petitions—as Apollo-Soyuz approached—insisting that NASA not prohibit acts of religious speech by the nation's astronauts, the space program was a source of acute spiritual satisfaction *and* disquiet. In its inspirations, in its import for notions of where man stood in relation to the divine, in the opportunities it presented for profound and novel experience, and in its role as an agent of transcendence within a secularizing state the program was implicated deeply in questions of ultimate concern. Within the wider culture of the long Sixties, the program was significant because it was religiously significant.

From one angle of view, the life-worlds inhabited by NASA, its installations, and its personnel seemed comfortably continuous with the exemplary landscapes of postwar American religiosity. His working week over, *homo astronautus*, like organization men everywhere, accompanied his family to services at the local suburban church. Surveying the national space program, Americans of faith were cheered to see many of their kind laboring dutifully there. Indeed, the commitment of NASA personnel to the project of making their country preeminent in space appeared almost to be modeled on the vocations of the pious. And when they spoke in religious terms about what they were doing, they could claim that their work conformed to a rich Christian symbolic tradition in which ascent affirmed the marriage between one's actions and the will of God.

Religious Americans, however, were also counted among the many millions who offered no strong support for the program. There were persuasive secular reasons to be ambivalent toward the country's enterprise in space, but those reasons came to seem persuasive in part because the program largely failed to articulate its purposes in noninstrumental terms and to make a case for exploration of the heavens that spoke to the actual spiritual commitments of Americans in the long Sixties. There was not much evidence that the inspirations of church and scripture had any significant purchase on the spaceflight policymaking process or NASA's operational routines. Nor did many space workers—engineers, managers, astronauts—make sense of their dedication by explicitly referring to a transcendent goal. NASA's standard, default invocation of an innate human will to explore was a poor substitute, for it reduced the objective of space travel to the service of an instinct and begged the question of how a purportedly universal attribute could produce only minority public approval for the agency's designs on the moon. And was there, moreover, some shadow of another, ungodly allegiance in NASA's reflexively secular, technocratic form of self-presentation, in its apparent reluctance to draw on the treasuries of spiritual language, ritual, and symbolism to anoint its ventures in spaceflight?

Across the Christian cosmological tradition, the space age threatened iconoclastic effects. Its rockets would puncture the sacred canopy, relegating heaven to the status of metaphor; man would proclaim his infinite potentialities, accelerating outward to encounters with other worlds. In the face of such anticipations, theologians remained sanguine. Indeed, they judged that the space age might actually aid their work, for it would force the laity (and some clergy too) to think anew about where they stood in relation to God. Initially, lay believers might be disoriented by the death of the old cosmography, but in time they would mature into a faith much better matched to the uncertainties of modern existence and a universe without end—a purer faith, stripped of extraneous metaphysics. The space age, however, did not turn out in quite the way that the theologians predicted. By the end of the long Sixties, many Americans were reinvesting their faith in the medieval cosmos, with earth at its center, a heaven in the skies, and mankind more than ever dependent on God. Spaceflight no longer promised a new doctrinal definition of the relation between man and the divine. Instead, in episodes such as the ill-fated flight of *Apollo 13*, the rule of the old relation had been dramatically reaffirmed. It was the assumption of a cosmic destiny for mankind, not the claims of conventional faith, that now seemed most open to doubt. The space age ended as an evangelical revival began.

Early in the space age, there also had been expansive speculations about whether spaceflight would be an occasion for spiritual experience. The media was curious to know whether, aloft in the heavens, the astronaut would feel closer to God. Space scientists conceived the question slightly differently: cut off from the earth, floating in an endless, silent void, would he begin to have visions? What fascinated many Americans about human spaceflight—the potentially transformative effects of a passage to the high places of sacred tradition or to an Archimedean position of vantage—was for NASA managers the stuff of nightmares. Some astronauts did return from their missions attesting to epiphany. Russell Schweickart, Edgar Mitchell, and James Irwin were so affected that they conceived of their experience as a gift to be shared with the world. In their efforts to explain what they had learned, these astronauts came closer than anyone else to realizing the prophetic potential of the manned spaceflight program. However, within the program itself they were regarded as misfits, precisely because in their encounters with the cosmos they had proven so susceptible to change. NASA placed a premium on stability in its astronauts, stigmatizing as eccentric and hazardous any behavior in space that went beyond a routine avowal of the beauty of the universe. An astronaut lost in wonder was not doing his job correctly and, in an environment unforgiving of error and neglect, might quickly become lost for good. The same insistent, self-conscious focus on operational priorities cramped the agency's style through the season of its greatest triumph: Neil Armstrong's first steps on the moon may have lifted the nation's mood, but the thrill of the moment was hardly likely to be quickened into a broader spiritual transformation by its coda of contingency samples, seismic experiments, and awkward repartee with Richard Nixon.

Indeed, the one occasion in the space age when NASA—*as* NASA—really muted the prattle of mission business and offered its audience an explicit prompt to spiritual reflection had already occurred, seven months before Armstrong and Aldrin landed on the moon. When the crew of *Apollo 8* addressed the world on Christmas Eve with a reading from Genesis, it showed what could happen when the cosmic perspective afforded by spaceflight was reconciled with a religious theme: a public paused in a common sensation of sacred time, thoughts deepening into ontology. But did the Genesis reading really express the core concerns of the space program? Did it make plain where the program stood in the contest between secularism and faith? Madalyn Murray O'Hair may have thought so, but many Christians were not sure. Prior to *Apollo 8*, pious observances within the program had been few and far between, and the right to prayer had already been surrendered in other arenas of state activity. Christians swamped NASA with letters and petitions, fearful that the agency might prohibit astronauts on future flights from speaking the name of God.

NASA responded with a murmur of acknowledgment. Its operational agnosticism held. Even though by the early 1970s the agency was seeking to cultivate new constituencies, it made little effort to allay the anxieties expressed by religious Americans and reengineer them into a basis for active support. In private, senior NASA officials were alert to the possibilities, but they ultimately did nothing, as if they too lacked conviction about how the agency's work corresponded with common conceptions of the divine. The result was a final disinvestment of religious hope in the space program. For many Americans of faith, as manned space operations paused after Apollo-Soyuz to await the development of the shuttle, spaceflight receded as a topic of immediate concern. The battle against secularism returned to earthly fronts—to the FCC, the IRS, and the liberal state writ large. The era of the shuttle, when it eventually arrived, seemed to confirm the point: it was a machine for routinizing low-orbit spaceflight, not for touching the face of God whether in awe or transgressive glee.

Now the shuttles themselves have been decommissioned, and the US manned space program is currently something of a hollow shell, a boilerplate mock-up without a prayer of a working rocket to deliver it into the heavens. In the Mercury-Apollo years, the program never conclusively reconciled the prophetic and the profane dimensions of spaceflight, but at least the question of their relation mattered. Here was a new technology of transcendence, engineered by a new breed of technocrats, flown into new environments by a new kind of man, under the aegis of a secularizing state, for purposes linked to the prosecution of a religious Cold War. The program simultaneously kindled thoughts of God and the death of God. When religion was absent from the program, its absence became a point of religious concern. When religion was present, its presence meant more, to more people, than it did in any other national undertaking of the time. Only for as long as that was true did Americans live in an age of space.

Notes

INTRODUCTION: The Blasphemy of Going Up

1. Oriana Fallaci, *If the Sun Dies* (New York: Atheneum, 1967), 379.
2. Freeman's fellow astronaut Pete Conrad, who was also friendly with Fallaci, later claimed that he had been the one to compare the launch to a prayer. See Santo L. Aricò, *Oriana Fallaci: The Woman and the Myth* (Carbondale: Southern Illinois University Press, 2010), 88.
3. Gene Kranz, *Failure Is Not an Option: Mission Control from Mercury to Apollo 13 and Beyond* (New York: Berkley Books, 2000), 122.
4. Fallaci, *If the Sun Dies*, vii.
5. Ibid., 20–21.
6. Ibid., 33.
7. Ibid., 95.
8. Ibid., 8.
9. Ibid., 379–80.
10. Ibid., 396.
11. John Morton Blum, *Years of Discord: American Politics and Society, 1961–1974* (New York: Norton, 1991); Allen J. Matusow, *The Unraveling of America: A History of Liberalism in the 1960s* (New York: Harper & Row, 1984); Maurice Isserman and Michael Kazin, *America Divided: The Civil War of the 1960s* (Oxford: Oxford University Press, 2000). On this point, see M. J. Heale, "The Sixties as History: A Review of the Political Historiography," *Reviews in American History* 33 (2005): 136–37.
12. See, e.g., Blum, *Years of Discord*.
13. Isserman and Kazin, *America Divided*, 263–64; Kim McQuaid, *The Anxious Years: America in the Vietnam-Watergate Era* (New York: Basic Books, 1989), 50.
14. Walter A. McDougall, *. . . the Heavens and the Earth: A Political History of the Space Age* (New York: Basic Books, 1985). For a critique of McDougall's thesis that *Sputnik 1* initiated a broad technocratic revolution as well as the space age, see Robert R. MacGregor, "Imagining an Aerospace Agency in the Atomic Age," in *Remembering the Space Age*, edited by Steven J. Dick (Washington, DC: NASA History Division, 2008), 55–70. MacGregor argues that in establishing NASA, policymakers were strongly influenced by precedents set in the creation of another technocratic institution, the Atomic Energy Commission, in the late 1940s.
15. McDougall, *. . . the Heavens and the Earth*, 451. For William E. Burrows, the closing of the "first space age" was contemporaneous with the end of the Cold War. However, the "second space age," which was both more militarized and more entrepreneur-

ial in character, began very quickly thereafter. Burrows, *This New Ocean: The Story of the First Space Age* (New York: Random House, 1998), 610–46.

16. Marina Benjamin, *Rocket Dreams: How the Space Age Shaped our Vision of a World Beyond* (London: Vintage, 2003), 17. See also Benjamin, "The End of the Space Age," *New Statesman*, 10 February 2003, www.newstatesman.com/200302100019.

17. For Walter McDougall, reflecting in 2007 on the fiftieth anniversary of *Sputnik 1*, mankind was still living in the space age, but he seemed to acknowledge that the term had lost much of its former salience. The anniversary, he observed, was "draped with a certain melancholy." McDougall asked his audience: "Do you sense a mood of disappointment, frustration, impatience over the failure of the human race to achieve much more than the minimum extrapolations made back in the 1950s, and considerably less than the buoyant expectations expressed as late as the 1970s?" Walter A. McDougall, "A Melancholic Space Age Anniversary," in Dick, *Remembering the Space Age*, 389.

18. Kim McQuaid, "Race, Gender, and Space Exploration: A Chapter in the History of the Space Age," *Journal of American Studies* 41 (2007): 405–34. For other useful discussions of women and the space program, see Martha Ackmann, *The Mercury 13: The True Story of Thirteen Women and the Dream of Space Flight* (New York: Random House, 2004); Bettyann Holtzmann Kevles, *Almost Heaven: The Story of Women in Space* (New York: Basic Books, 2003); Daniel Sage, "Giant Leaps and Forgotten Steps: NASA and the Performance of Gender," in *Space Travel and Culture: From Apollo to Space Tourism*, edited by David Bell and Martin Parker (Oxford: Wiley-Blackwell, 2009), 146–63; David J. Shayler and Ian Moule, *Women in Space: Following Valentina* (New York: Springer, 2005), 69–109; and Margaret A. Weitekamp, *Right Stuff, Wrong Sex: America's First Women in Space Program* (Baltimore: Johns Hopkins University Press, 2004). On race, see Andrew J. Dunar and Stephen P. Waring, *Power to Explore: A History of Marshall Space Flight Center, 1960–1990* (Washington, DC: NASA, 1999), 115–25; and J. Alfred Phelps, *They Had a Dream: The Story of African-American Astronauts* (Novato, CA: Presidio, 1994).

19. Kim McQuaid, "Selling the Space Age: NASA and Earth's Environment, 1958–1990," *Environment and History* 12 (2006): 127–63.

20. Alan Bean, oral history interview by Michelle Kelly, 23 June 1998, NASA/Johnson Space Center (JSC) Oral History Project, 22, www.jsc.nasa.gov/history/oral_histories/oral_histories.htm.

21. Howard E. McCurdy, *Space and the American Imagination* (Washington, DC: Smithsonian Institution Press, 1997); Robert Poole, *Earthrise: How Man First Saw the Earth* (New Haven, CT: Yale University Press, 2008). Steven J. Dick and Roger Launius refer to a "new aerospace history" in their introduction to *Critical Issues in the History of Spaceflight*, edited by Dick and Launius (Washington, DC: NASA History Division, 2006), vii–xi. See also Glen Asner, "Space History from the Bottom Up: Using Social History to Interpret the Societal Impact of Spaceflight," in *Societal Impact of Spaceflight*, edited by Steven J. Dick and Roger D. Launius (Washington, DC: NASA History Division, 2007), 387–405; Roger D. Launius, "The Historical Dimension of Space Exploration: Reflections and Possibilities," *Space Policy* 16 (2000): 23–38; Asif A. Siddiqi, "American Space History: Legacies, Questions, and Opportunities for Future Research," in Dick and Launius, *Critical Issues*, 433–80; and Margaret A. Weitekamp, "Critical Theory as a Toolbox: Suggestions for Space History's Relationship to the History Subdisciplines," in ibid., 549–72.

22. This was the poignant fate of Wernher von Braun, for long years the most persuasive salesman of the space program and the potential for applying its technology and management methods to the problems of society at large. Michael J. Neufeld, *Von Braun: Dreamer of Space, Engineer of War* (New York: Knopf, 2007), 434–72.

23. Important recent works on postwar religion include Darren Dochuk, *From Bible Belt to Sun Belt: Plain-Folk Religion, Grassroots Politics and the Rise of Evangelical Conservatism* (New York: Norton, 2011); Steven P. Miller, *Billy Graham and the Rise of the Republican South* (Philadelphia: University of Pennsylvania Press, 2009); Bethany Moreton, *To Serve God and War-Mart: The Making of Christian Free Enterprise* (Cambridge, MA: Harvard University Press, 2009); Jeff Sharlet, *The Family: The Secret Fundamentalism at the Heart of American Power* (New York: Harper Perennial, 2008); John G. Turner, *Bill Bright and Campus Crusade for Christ: The Renewal of Evangelicalism in Postwar America* (Chapel Hill: University of North Carolina Press, 2008); and Daniel Williams, *God's Own Party: The Making of the Christian Right* (Oxford: Oxford University Press, 2010).

24. Jon Butler, "Jack-in-the-Box Faith: The Religion Problem in Modern American History," *Journal of American History* 90 (2004): 1357–78.

25. See, e.g., Burrows, *This New Ocean*; and Andrew Chaikin, *A Man on the Moon: The Voyages of the Apollo Astronauts* (London: Penguin, 1998).

26. Poole, *Earthrise*, 105–8, 128–40.

27. There is no discussion of space exploration, for example, in Amanda Porterfield's otherwise wide-ranging book *The Transformation of American Religion: The Story of a Late Twentieth-Century Awakening* (Oxford: Oxford University Press, 2001). The topic receives incidental attention in Robert S. Ellwood, *The Sixties Spiritual Awakening: American Religion Moving from Modern to Postmodern* (New Brunswick, NJ: Rutgers University Press, 1994).

28. Ellwood, *Sixties Spiritual Awakening*, 48–49, 303, 305–11.

29. David F. Noble, *The Religion of Technology: The Divinity of Man and the Spirit of Invention* (New York: Knopf, 1997).

CHAPTER ONE: A Power Greater Than Any of Us

1. Gene Farmer and Dora Jane Hamblin, *First on the Moon: A Voyage with Neil Armstrong, Michael Collins and Edwin E. Aldrin, Jr.* (Boston: Little, Brown, 1970), 267–68.

2. The quotations in Woodruff's paper are from Mircea Eliade, *Myths, Dreams, and Mysteries: The Encounter between Contemporary Faiths and Archaic Realities*, translated by Philip Mairet (1960; reprint, New York: Harper & Row, 1975), 103–6.

3. Farmer and Hamblin, *First on the Moon*, 268.

4. Eliade, *Myths, Dreams, and Mysteries*, 23–35.

5. Mircea Eliade, *Images and Symbols: Studies in Religious Symbolism* (Kansas City: Sheed Andrews & McMeel, 1961), 18–21; Eliade, *The Myth of the Eternal Return: Cosmos and History*, 2nd ed. (Princeton, NJ: Princeton University Press, 2005), 141–62.

6. No interest in spaceflight is evident in Mircea Eliade, *No Souvenirs: Journal, 1957–1969* (London: Routledge & Kegan Paul, 1978).

7. Eliade, *Images and Symbols*, 166–67; Eliade, *Myths, Dreams, and Mysteries*, 66–72.

8. Noble, *Religion of Technology*, 5. See also Ryan J. McMillen, "Space Rapture: Extraterrestrial Millennialism and the Cultural Construction of Space Colonization" (PhD diss., University of Texas at Austin, 2004).

9. The basis for a more sophisticated comprehension of human motivation may be found in David C. McClelland, *Human Motivation* (Cambridge: Cambridge University Press, 1985).

10. For a valuable discussion of sociological debates about modernity and secularization, see Roy Wallis and Steve Bruce, "Secularization: The Orthodox Model," in *Religion and Modernization: Sociologists and Historians Debate the Secularization Thesis*, edited by Steve Bruce (Oxford: Oxford University Press, 1992), 8–30.

11. For evidence of both the evangelical revival and the wider persistence of religious belief, see "The Christianity Today–Gallup Poll: An Overview," *Christianity Today*, 21 December 1979, 12–15. For an example of how a theorist of secularization eventually changed his mind, compare Peter Berger, *The Sacred Canopy: Elements of a Sociological Theory of Religion* (1967; reprint, New York: Anchor Books, 1990), with Berger, "The Desecularization of the World: A Global Overview," in *The Desecularization of the World: Resurgent Religion and World Politics*, edited by Peter Berger (Grand Rapids, MI: Wm. B. Eerdmans, 1999), 1–18.

12. See, most notably, Gerhard Lenski, *The Religious Factor: A Sociologist's Inquiry* (Garden City, NY: Anchor Books, 1963); and Robert Wuthnow, *God and Mammon in America* (New York: Free Press, 1994). On the general failure of American social scientists since Lenski to inquire into the broader social impacts of religious belief, see Craig Calhoun, "Gerhard Lenski: Some False Oppositions, and 'The Religious Factor,'" *Sociological Theory* 22 (2004): 194–204. A comprehensive survey of the consequences of religious commitment was planned in the late 1960s by Rodney Stark and Charles Y. Glock, but it seems never to have been completed. See Stark and Glock, *American Piety: The Nature of Religious Commitment* (Berkeley: University of California Press, 1968), 5–6.

13. See, e.g., Benjamin, *Rocket Dreams*, 57.

14. Emile Durkheim, *The Elementary Forms of the Religious Life*, 2nd ed. (London: George Allen & Unwin, 1976).

15. Berger, *Sacred Canopy*, 133–34.

16. Bronislaw Malinowski, *Magic, Science, and Religion, and Other Essays* (Boston: Beacon, 1948), 18. See also Clifford Geertz, "Religion as a Cultural System," in *The Interpretation of Cultures: Selected Essays* (New York: Basic Books, 1973), 119–20.

17. Malinowski, *Magic, Science, and Religion*, 40.

18. Durkheim, *Elementary Forms*, 37. For a similar insight, see Mircea Eliade, *The Sacred and the Profane: The Nature of Religion* (New York: Harcourt, 1957).

19. Geertz, "Religion as a Cultural System," 119–20.

20. Noble, *Religion of Technology*, 4–5.

21. Max Weber, *The Protestant Ethic and the Spirit of Capitalism* (Mineola, NY: Dover, 2003), 181.

22. See Andrew Preston, "Bridging the Gap between the Sacred and the Secular in the History of American Foreign Relations," *Diplomatic History* 30 (2006): 809–11.

23. See Butler, "Jack-in-the-Box Faith."

24. Saul Bellow, *Mr. Sammler's Planet* (Harmondsworth, UK: Penguin, 1972), 134, 217. For a valuable discussion of this novel, see William D. Atwill, *Fire and Power: The American Space Program as Postmodern Narrative* (Athens: University of Georgia Press, 1994), 25–44.

25. Norman Mailer, *A Fire on the Moon* (London: Weidenfeld & Nicolson, 1970), 379–80.

26. Lynn White Jr., "Dynamo and Virgin Reconsidered," in *Machina Ex Deo: Essays in the Dynamism of Western Culture* (Cambridge, MA: MIT Press, 1968), 57–73.

27. Lynn White Jr., "The Historical Roots of Our Ecologic Crisis," in ibid., 85–90; Noble, *Religion of Technology*, 12–18.

28. Noble, *Religion of Technology*, 90–91; Perry Miller, *The Life of the Mind in America: From the Revolution to the Civil War* (New York: Harcourt, Brace & World, 1965), 52, 90–91; David E. Nye, *America as Second Creation: Technology and Narratives of New Beginning* (Cambridge, MA: MIT Press, 2003).

29. Stephen Ellington, *The Megachurch and the Mainline: Rethinking Religious Tradition in the Twenty-first Century* (Chicago: University of Chicago Press, 2007), 42; Quentin J. Schultze, ed., *American Evangelicals and the Mass Media* (Grand Rapids, MI: Academie, 1990); Charles H. Lippy, *Do Real Men Pray? Images of the Christian Man and Male Spirituality in White Protestant America* (Knoxville: University of Tennessee Press, 2005), 207–20.

30. John F. Kasson, *Civilizing the Machine: Technology and Republican Values in America, 1776–1900* (Harmondsworth, UK: Penguin, 1977), 186–88.

31. Henry George, *Progress and Poverty* (New York: Cosimo, 2005), 11.

32. Jay Newman, *Religion and Technology: A Study in the Philosophy of Culture* (Westport, CT: Praeger, 1997), 9–23.

33. Peter Berger, *A Rumour of Angels: Modern Society and the Rediscovery of the Supernatural* (London: Allen Lane, 1970), 60–62.

34. Robert Perucci and Joel E. Gerstl, *Profession Without Community: Engineers in American Society* (New York: Random House, 1969); Robert Perucci and Joel E. Gerstl, eds., *The Engineers and the Social System* (New York: John Wiley & Sons, 1969); Edwin T. Layton Jr., *The Revolt of the Engineers: Social Responsibility and the American Engineering Profession* (Cleveland, OH: Press of Case Western Reserve University, 1971).

35. Samuel C. Florman, *The Existential Pleasures of Engineering* (London: Souvenir, 1995), 137; Robert M. Pirsig, *Zen and the Art of Motorcycle Maintenance: An Inquiry into Values* (New York: Bantam Books, 1984).

36. Eliade, *Myths, Dreams, and Mysteries*, 59–60, 98–107.

37. Eliade, *Images and Symbols*, 160–72.

38. See Richard P. Hallion, *Taking Flight: Inventing the Aerial Age from Antiquity through the First World War* (Oxford: Oxford University Press, 2003), 11–16.

39. Lynn White Jr., "Eilmer of Malmesbury, An Eleventh Century Aviator: A Case Study of Technological Innovation, Its Context and Tradition," in *Medieval Religion and Technology: Collected Essays* (Berkeley: University of California Press, 1978), 59–73.

40. Laurence Goldstein, *The Flying Machine and Modern Literature* (Bloomington: Indiana University Press, 1986), 14–40.

41. Hallion, *Taking Flight*, 7.

42. Paul Hoffman, *Wings of Madness: Alberto Santos-Dumont and the Invention of Flight* (London: Fourth Estate, 2003), 4, 152–53; Joseph J. Corn, *The Winged Gospel: America's Romance with Aviation* (Baltimore: Johns Hopkins University Press, 2002), 37–38; Michael S. Sherry, *The Rise of American Air Power: The Creation of Armageddon* (New Haven, CT: Yale University Press, 1987), 5–6.

43. Corn, *Winged Gospel*, 44–46; Sherry, *Rise of American Air Power*, 19–21.

44. Antoine de Saint-Exupéry, *Wind, Sand and Stars*, translated by William Rees (1995; reprint, London: Penguin, 2000), originally published in French as *Terre des hommes* (Paris: Gallimard, 1939); Stacy Schiff, *Saint-Exupéry: A Biography* (London: Chatto & Windus, 1994), 309–12.

45. Corn, *Winged Gospel*, 43–44.

46. James Tobin, *First to Fly: The Unlikely Triumph of Wilbur and Orville Wright* (London: John Murray, 2003), 43–44.

47. Ibid., 49–51; Tom D. Crouch, *The Bishop's Boys: A Life of Wilbur and Orville Wright* (New York: Norton, 1989), 159–64; Fred Howard, *Wilbur and Orville: A Biography of the Wright Brothers* (London: Robert Hale, 1988), 13–14, 28–32; Peter L. Jakab, *Visions of a Flying Machine: The Wright Brothers and the Process of Invention* (Shrewsbury, UK: Airlife, 1990).

48. Crouch, *Bishop's Boys*, 300.

49. Robert Wohl, *A Passion for Wings: Aviation and the Western Imagination, 1908–1918* (New Haven, CT: Yale University Press, 1994), 29–30.

50. Charles A. Lindbergh, *The Spirit of St. Louis* (New York: Charles Scribner's Sons, 1955), 13–16.

51. On responses to Lindbergh's flight, see Modris Eksteins, *Rites of Spring: The Great War and the Birth of the Modern Age* (New York: Bantam Books, 1989), 242–74; and Robert Wohl, *The Spectacle of Flight: Aviation and the Western Imagination, 1920–1950* (New Haven, CT: Yale University Press, 2005), 9–46.

52. Myron T. Herrick, quoted in Fitzhugh Green, "What the World Thought of Lindbergh," in *We—Pilot and Plane*, by Charles A. Lindbergh (London: G. P. Putnam's Sons, 1927), 185.

53. Charles A. Lindbergh, *Of Flight and Life* (New York: Charles Scribner's Sons, 1948), 49–51; Charles A. Lindbergh, *Autobiography of Values* (New York: Harcourt Brace Jovanovich, 1977), 5–6, 308, 384.

54. Charles A. Lindbergh, *Autobiography of Values*, 15–16.

55. The phrase is taken from the title of Corn's book, *The Winged Gospel*.

56. See esp. Corn, *Winged Gospel*.

57. On fascism as a "political religion," see Roger Griffin, ed., *Fascism, Totalitarianism and Political Religion* (Abingdon, UK: Routledge, 2005).

58. Wohl, *Spectacle of Flight*, 49–106.

59. *Triumph of the Will*, directed by Leni Riefenstahl (1935; London, UK: Simply Media, 2010), DVD; Wohl, *Spectacle of Flight*, 306.

60. A. Scott Berg, *Lindbergh* (London: Macmillan, 1998), 361, 377–80.

61. See esp. Charles A. Lindbergh, *Of Flight and Life*.

62. Corn, *Winged Gospel*, 66.

63. Col. Robert L. Scott, *God Is My Co-Pilot* (New York: Ballantine Books, 1971), ix.

64. See, e.g., Ernest K. Gann, *Fate Is the Hunter* (New York: Touchstone, 1961), 1–14; and General Chuck Yeager and Leo Janos, *Yeager: An Autobiography* (London: Century, 1986), 318–19.

65. Tom Wolfe, *The Right Stuff* (1989; reprint, London: Black Swan, 1990), 29–30, 37–39, 61–62; first published in 1979.

66. McDougall, . . . *the Heavens and the Earth*, 4.

67. Noble, *Religion of Technology*, 120–29. See also McMillen, "Space Rapture."

68. Burrows, *This New Ocean*, 36–44.

69. William Sims Bainbridge, *The Spaceflight Revolution: A Sociological Study* (New York: John Wiley & Sons, 1976), 31.

70. George Pendle, *Strange Angel: The Otherworldly Life of Rocket Scientist John Whiteside Parsons* (London: Phoenix, 2006).

71. Mailer, *Fire on the Moon*, 137–38.

72. McCurdy, *Space and the American Imagination*, 13–16.

73. William Butcher, *Jules Verne: The Definitive Biography* (New York: Thunder's Mouth, 2006), 281–82.

74. David A. Clary, *Rocket Man: Robert H. Goddard and the Birth of the Space Age* (New York: Theia, 2003), 258.

75. M. G. Lord, *Astro Turf: The Private Life of Rocket Science* (New York: Walker, 2005), 71.

76. Burrows, *This New Ocean*, 42–43.

77. Bainbridge, *Spaceflight Revolution*, 31.

78. Pendle, *Strange Angel*, 236–41.

79. Asif A. Siddiqi, *Challenge to Apollo: The Soviet Union and the Space Race, 1945–1974* (Washington, DC: NASA, 2000), 6–9; Neufeld, *Von Braun*, 49–88.

80. Siddiqi, *Challenge to Apollo*, 11–16; Neufeld, *Von Braun*, 169–73.

81. Neufeld, *Von Braun*, 21–37, 152.

82. Wernher von Braun to editor, *Christian Century*, 27 January 1960, 106.

83. On von Braun's conversion, see Neufeld, *Von Braun*, 229–30.

84. Daniel Lang, *From Hiroshima to the Moon: Chronicles of Life in the Atomic Age* (New York: Dell, 1961), 218; Wernher von Braun, "A Scientist's Belief in God," address to Governor's Prayer Breakfast, Denver, Colorado, 1 April 1969, folder: "Speeches and Writings File: April–August 1969," box 49, Wernher von Braun Papers, Manuscript Division, Library of Congress, Washington, DC.

85. Neufeld, *Von Braun*, 477.

86. For a useful survey of the revival, see James Hudnut-Beumler, *Looking for God in the Suburbs: The Religion of the American Dream and Its Critics, 1945–1965* (New Brunswick, NJ: Rutgers University Press, 1994), 29–84.

87. Joel A. Carpenter, *Revive Us Again: The Reawakening of American Fundamentalism* (Oxford: Oxford University Press, 1997), 211–32.

88. See, e.g., Gallup poll, 21 April 1957, in *The Gallup Poll: Public Opinion, 1935–1971*, by George H. Gallup, 3 vols. (New York: Random House, 1972), vol. 2, *1949–1958*, 1482. On Billy Graham's courtship of political leaders, see William Martin, *With God on Our Side: The Rise of the Religious Right in America* (New York: Broadway Books, 2005), 30–33, 40–42.

89. Dianne Kirby, "Harry Truman's Religious Legacy: The Holy Alliance, Containment and the Cold War," in *Religion and the Cold War*, edited by Dianne Kirby (Basingstoke, UK: Palgrave Macmillan, 2003), 77–98; William Inboden, *Religion and American Foreign Policy, 1945–1960: The Soul of Containment* (New York: Cambridge University Press, 2008).

90. Robert S. Ellwood, *The Fifties Spiritual Marketplace: American Religion in a Decade of Conflict* (New Brunswick, NJ: Rutgers University Press, 1997), 38–43.

91. T. Jeremy Gunn, *Spiritual Weapons: The Cold War and the Forging of an American National Religion* (Westport, CT: Praeger, 2009), 64–69.

92. Anne C. Loveland, *American Evangelicals and the U.S. Military, 1942–1993* (Baton Rouge: Louisiana State University Press, 1996), 10–14.

93. Bruce J. Dierenfield, *The Battle over School Prayer: How* Engel v. Vitale *Changed America* (Lawrence: University Press of Kansas, 2007), 66.

94. Susan Jacoby, *Freethinkers: A History of American Secularism* (New York: Metropolitan, 2004), 291, 309–12.

95. See, e.g., Carl F. H. Henry, "Can We Salvage the Republic?," *Christianity Today*, 3 March 1958, 3–7.

96. Billy Graham to Dwight D. Eisenhower, 2 December 1957, Record No. 6774, NASA History Office, Washington, DC.

97. Henry, "Can We Salvage the Republic?"; "Public Funds for Public Schools," *Christianity Today*, 10 April 1961, 20–23.

98. Michael L. Smith, "Selling the Moon: The U.S. Manned Space Program and the Triumph of Commodity Scientism," in *The Culture of Consumption: Critical Essays in American History, 1880–1980*, edited by Richard Wightman Fox and T. J. Jackson Lears (New York: Pantheon, 1983), 194.

99. Linda T. Krug, *Presidential Perspectives on Space Exploration: Guiding Metaphors from Eisenhower to Bush* (New York: Praeger, 1991), 23–30; McDougall, . . . *the Heavens and the Earth*, 141–76.

100. See, e.g., the transcript of a discussion between Kennedy and NASA administrator James Webb, 21 November 1962, University of Virginia Miller Center Presidential Recordings Program, http://tapes.millercenter.virginia.edu/clips/1962_1121_apollo/.

101. McDougall, . . . *the Heavens and the Earth*, 315–24.

102. See, e.g., President's Science Advisory Committee, "Introduction to Outer Space," 26 March 1958, in *Sputnik, Scientists, and Eisenhower: A Memoir of the First Special Assistant to the President for Science and Technology*, by James R. Killian Jr. (Cambridge, MA: MIT Press, 1977), 288–99. The statement lists "four factors which give importance, urgency, and inevitability to the advancement of space technology," the first being "the compelling urge of man to explore and to discover" and the third, "national prestige."

103. Michael L. Smith, "Selling the Moon," 195.

104. For a detailed discussion of the language and images used by NASA to sell the space program, see Mark E. Byrnes, *Politics and Space: Image Making by NASA* (Westport, CT: Praeger, 1994).

105. "Statement of Preliminary U.S. Policy on Outer Space," 18 August 1958, NSC 5814/1, in *Foreign Relations of the United States, 1958–1960*, vol. 2, *United Nations and General International Matters*, edited by Suzanne E. Coffman and Charles S. Sampson (Washington, DC: Government Printing Office, 1991), 845–63.

106. John F. Kennedy, "Remarks at the 10th Annual Presidential Prayer Breakfast," 1 March 1962, *Public Papers of the Presidents: John F. Kennedy, 1962* (Washington, DC: Government Printing Office, 1963), 175–76.

107. "We are extending this God-given brain and these God-given hands to their outermost limits," von Braun told an audience in Huntsville after the safe return of the *Apollo 11* astronauts. "Someday the universe will be ours," declared the draft of another address. A scrawled annotation amended it thus: "Someday our little corner of the universe will be ours." Draft of speeches at an afternoon picnic and evening

dinner held in celebration of the lunar landing, 26 July 1969, folder: "Speeches and Writings File: April–August 1969," box 49, von Braun Papers.

108. Von Braun, "Scientist's Belief in God"; Erik Bergaust, *Wernher Von Braun* (Washington, DC: National Space Institute, 1976), 109–17.

109. Hugh L. Dryden, "The Importance of Religion in American Life," 12 March 1950, in *The Impact of Space on Religion*, edited by Martha Wheeler George, NASA Historical Note No. 25 (Washington DC: NASA, 1963).

110. Gordon Harris, *A New Command: The Life of Bruce Medaris, Major General USA, Retired* (Plainfield, NJ: Logos International, 1976).

111. Kranz, *Failure Is Not an Option*, 101–15, 277. The postwar vagaries of military manpower requirements also played a role: Kranz applied to NASA in 1960 only after the air force had turned down his application for a return to active duty.

112. T. Keith Glennan, *The Birth of NASA: The Diary of T. Keith Glennan* (Washington, DC: NASA, 1993); Piers Bizony, *The Man Who Ran the Moon: James Webb, JFK and the Secret History of Project Apollo* (Cambridge: Icon Books, 2006); W. Henry Lambright, *Powering Apollo: James E. Webb of NASA* (Baltimore: Johns Hopkins University Press, 1995).

113. Bizony, *Man Who Ran the Moon*, 20–22; Lambright, *Powering Apollo*, 6, 82–87.

114. See Robert R. Gilruth oral history interviews, 1986–87, Glennan-Webb-Seamans Project for Research in Space History, Space History Division, National Air and Space Museum, Washington, DC, www.nasm.si.edu/research/dsh/gwspi-p1.html.

115. Chris Kraft, *Flight: My Life in Mission Control* (New York: Plume, 2002), 235.

116. Wesley Hjornevik, "MSC Announcement No. 183: Religious Holidays," 9 April 1963, folder: "#95-194," box 3, Issuances 1961–1966, Records of the Johnson Space Center, RG 255, Records of the National Aeronautics and Space Administration, National Archives Southwest Region, Fort Worth, TX; Robert Gilruth to George Mueller, "GT-6/GT-7 crew post-flight schedules," 19 November 1965, folder: "Gemini," box 3, Flight Crew Operations Directorate, Director's Subject Files, 1964–1970, Records of the Johnson Space Center.

117. Deke Slayton, affidavit, 3 October 1969, "Civil 3502* (Madalyn O'Hair)," Records of US Attorney, Western District of Texas, San Antonio Division, RG 118, Records of United States Attorneys and Marshals, National Archives Southwest Region, Fort Worth.

118. Gordon Cooper, *Leap of Faith: An Astronaut's Journey into the Unknown*, with Bruce Henderson (New York: HarperCollins, 2000), 67–68.

119. Frank Borman, "Message to Earth," *Guideposts*, April 1969, 4.

120. Kenneth Atchison to Belton Currington, 26 July 1966, Record No. 6774, NASA History Office.

121. *New York Times*, 28 December 1968, 14; Thomas Paine, notes of talk to the Arkansas Brotherhood of the National Conference of Christians and Jews, 5 May 1969, folder: "Reading File: May 1969," 1 of 2, box 33, Government File, Thomas O. Paine Papers, Manuscript Division, Library of Congress, Washington, DC.

122. "Space, Science, and Scripture," *Christianity Today*, 18 July 1969, 3–6.

123. Mailer, *Fire on the Moon*.

124. Lord, *Astro Turf*, 8.

125. Sylvia D. Fries, *NASA Engineers and the Age of Apollo* (Washington, DC: NASA, 1992).

126. Howard E. McCurdy, *Inside NASA: High Technology and Organization Change in the U.S. Space Program* (Baltimore: Johns Hopkins University Press, 1993), 82–83.

127. Mailer, *Fire on the Moon*, 10; Fallaci, *If the Sun Dies*, 161.

128. Mailer, *Fire on the Moon*, 49.

129. Michael Collins, *Carrying the Fire: An Astronaut's Journey* (New York: Bantam Books, 1983), 349.

130. Kranz, *Failure is Not an Option*, 136, 140, 253.

131. Ibid., 177.

132. Kevin M. Brady, "NASA Launches Houston into Orbit: The Economic and Social Impact of the Space Agency on Southeast Texas, 1961–1969," in Dick and Launius, *Societal Impact of Spaceflight*, 451–65; Henry C. Dethloff, *Suddenly Tomorrow Came: A History of the Johnson Space Center* (Houston: NASA, 1993), 146–52, 257–60; Dunar and Waring, *Power to Explore*, 59–64; William Barnaby Faherty, *Florida's Space Coast: The Impact of NASA on the Sunshine State* (Gainesville: University Press of Florida, 2002), 53–58, 115–18.

133. Ann Markusen, Peter Hall, Scott Campbell, and Sabrina Deitrick, *The Rise of the Gunbelt: The Military Remapping of Industrial America* (New York: Oxford University Press, 1991).

134. Faherty, *Florida's Space Coast*, 57.

135. Annie Mary Hartsfield, Mary Alice Griffin, and Charles M. Grigg, *Summary Report: NASA Impact on Brevard County* (Tallahassee: Institute for Social Research, Florida State University, 1966), 10.

136. Dethloff, *Suddenly Tomorrow Came*, 211, 214; Dunar and Waring, *Power to Explore*, 151–52; Faherty, *Florida's Space Coast*, 112; David McComb, *Houston: A History* (Austin: University of Texas Press, 1981), 143–44.

137. Etan Diamond, *Souls of the City: Religion and the Search for Community in Postwar America* (Bloomington: Indiana University Press, 2003), 24–32.

138. For the classic statement of this argument, see Gibson Winter, *The Suburban Captivity of the Churches: An Analysis of Protestant Responsibility in the Expanding Metropolis* (Garden City, NY: Doubleday, 1961). A similar case was made in William H. Whyte, *The Organization Man* (Harmondsworth, UK: Penguin, 1960), 337–51. For a discussion of the suburban jeremiad, see Hudnut-Beumler, *Looking for God*, 131–66.

139. Elise Hopkins Stephens, *Historic Huntsville: A City of New Beginnings* (Sun Valley, CA: American Historical Press, 2002), 92.

140. Ibid., 106–10.

141. Neufeld, *Von Braun*, 217–25, 238–39.

142. Ibid., 245, 248–49.

143. Dunar and Waring, *Power to Explore*, 15–16; Stephens, *Historic Huntsville*, 117.

144. Dunar and Waring, *Power to Explore*, 126–29.

145. See the website of St. Mark's Evangelical Lutheran Church, Huntsville, AL, www.stmarkshsv.org/.

146. Most notoriously, in 1961 Marshall engineers demanded an additional test of the Mercury-Redstone rocket before it was used to place a man in space. As a result, Yuri Gagarin got there first. Neufeld, *Von Braun*, 358–59.

147. A lively personal account of this revival is offered in Lucile Johnston, *Will We Find Our Way? A Space-Age Odyssey* (Atlanta: Cross Roads Books, 1979).

148. Noble, *Religion of Technology*, 128–30; "NASA's Lucas: Scientist and Baptist Lay

Leader," undated article enclosed in James C. Fletcher to William R. Lucas, 7 April 1975, folder 1: "Correspondence (April 1975)," box 34, James C. Fletcher Papers, Special Collections, J. Willard Marriott Library, University of Utah, Salt Lake City.

149. Dethloff, *Suddenly Tomorrow Came*, 97–99; James R. Hansen, *Engineer in Charge: A History of the Langley Aeronautical Laboratory, 1917–1958* (Washington, DC: NASA, 1987).

150. Dethloff, *Suddenly Tomorrow Came*, 165, 169.

151. Fallaci, *If the Sun Dies*, 267.

152. Mailer, *Fire on the Moon*, 100.

153. Dethloff, *Suddenly Tomorrow Came*, 65, 68.

154. Buzz Aldrin, "Communion in Space," *Guideposts*, October 1970, 1, 3–6; Kranz, *Failure Is Not an Option*, 178; *Washington Evening Star*, 7 August 1971, A-6; *New York Times*, 30 December 1968, 1, 18; James R. Hansen, *First Man: The Life of Neil Armstrong; The First Authorised Biography* (London: Simon & Schuster, 2005), 477–78.

155. See Apollo Prayer League newsletter, c. February 1971, Record No. 6744, NASA History Office.

156. Aldrin, "Communion in Space," 1, 3–6.

157. Faherty, *Florida's Space Coast*, 2.

158. Frank H. Thomas Jr., *Day by Day: A Pilgrimage in Faith; A History of First Baptist Church, Cocoa, Florida, 1910–1990* (Cocoa: First Baptist Church of Cocoa, 1990).

159. "History of Riverside Presbyterian Church, Cocoa Beach, Florida," Riverside Presbyterian Church website, www.riversidepres.org.

160. Guenter Wendt and Russell Still, *The Unbroken Chain* (Burlington, ON: Apogee, 2001), 16.

161. Wolfe, *Right Stuff*, 258; John Glenn, *John Glenn: A Memoir*, with Nick Taylor (New York: Bantam Books, 1999), 245.

162. Hartsfield, Griffin, and Grigg, *Summary Report*, 87.

163. Thomas, *Day by Day*, 228–32.

164. For a detailed account of the construction of the Apollo launch complex, see Charles D. Benson and William Barnaby Faherty, *Moonport: A History of Apollo Launch Facilities and Operations* (Washington, DC: NASA, 1978).

165. Thomas, *Day by Day*, 247.

166. Wolfe, *Right Stuff*, 145–46; Neal Thompson, *Light This Candle: The Life and Times of Alan Shepard* (New York: Three Rivers, 2005), 242–46.

167. "Life in the Space Age," *Time*, 4 July 1969.

168. Robert L. Lowry, "Recollections," Riverside Presbyterian Church website, www.riversidepres.org.

169. "History of Riverside Presbyterian Church"; Penny Dale and John Dale, "Recollections," Riverside Presbyterian Church website.

170. *Christian Beacon*, 7 January 1971, 2, 5, 8, and 28 January 1971, 1; Donald Janson, "Right-Wing Cleric May Face Eviction," *New York Times*, 23 August 1974, 36.

171. Janson, "Right-Wing Cleric"; "Carl McIntire: On the Move from Cape to Cape," *Christianity Today*, 17 August 1979, 45–46.

172. Janson, "Right-Wing Cleric."

173. Victor C. Ferkiss, *Technological Man: The Myth and the Reality* (London: Heinemann, 1969), 3–7.

174. Mailer, *Fire on the Moon*, 223.

175. Fallaci, *If the Sun Dies*, 307–8.

176. Frank Gannon to Ken Cole, "The Space Program: the President's Domestic Policy and the President's Domestic Image," 22 May 1973, enclosed in Fletcher to Howard E. McCurdy, 8 November 1973, folder 3: "Correspondence (8 Nov–28 Dec 1973)," box 17, Fletcher Papers.

177. Ferkiss, *Technological Man*, 5.

178. For an example of such a judgment, see Kraft, *Flight*, 164–70. For Carpenter's defense of his performance, see Scott Carpenter and Kris Stoever, *For Spacious Skies: The Uncommon Journey of a Mercury Astronaut* (New York: New American Library, 2004), 256–303.

179. Collins, *Carrying the Fire*, 52.

180. Chaikin, *Man on the Moon*, 47–49.

181. "Press Conference: Mercury Astronaut Team," 9 April 1959, folder: "April 1–16, 1959," box 3, Chronological Files, NASA History Office Source Files on Project Mercury, National Archives Southwest Region, Fort Worth.

182. Wolfe, *Right Stuff*, 102–7.

183. Ibid., 144–47.

184. "Press Conference: Mercury Astronaut Team."

185. M. Scott Carpenter, L. Gordon Cooper Jr., John H. Glenn Jr., Virgil I. Grissom, Walter M. Schirra Jr., Alan B. Shepard Jr., and Donald K. Slayton, *We Seven* (New York: Simon & Schuster, 1962), 13. See also Glenn, *John Glenn*, 182–83.

186. Carpenter et al., *We Seven*, 33–34.

187. Glenn, *John Glenn*, 179.

188. For Pete Conrad, the motivation was flying; for Jim Lovell, rocketry; for William Anders, exploration. In 1965 NASA recruited five scientist-astronauts, including the geologist Harrison Schmidt, who traveled to the moon seven years later aboard *Apollo 17*. See Nancy Conrad and Howard A. Klausner, *Rocket Man: Astronaut Pete Conrad's Incredible Ride to the Moon and Beyond* (New York: New American Library, 2005), 107; Jim Lovell and Jeffrey Kluger, *Apollo 13* (New York: Pocket Books, 1995), 65; and Chaikin, *Man on the Moon*, 39, 386.

189. *La Mesa Scout*, 2 January 1969, enclosed in Robert Sherrod to John McLeaish, 24 January 1969, Record No. 013130, folder: "Anders, Major William A., USAF—Apollo 8 Crewman," Robert Sherrod Collection, NASA History Office; Frank Borman, *Countdown: An Autobiography*, with Robert J. Serling (New York: Silver Arrow Books, 1988), 79, 197–98; Collins, *Carrying the Fire*, 58.

190. On Borman, see Borman, "Message to Earth," 3; and Homer Bigart, "New Breed Astronauts: Scientists, Not Daredevils," *New York Times*, 30 December 1968, 18. On White, see "Space: To Strive, To Seek, To Find, And Not to Yield . . . ," *Time*, 3 February 1967, www.time.com; and Graham Purcell, "The Faith of Astronaut Edward H. White II," Cong. Rec. 19465–66 (1969). On Scott, see McCandlish Phillips, "Family Portraits of 3 Astronauts," *New York Times*, 15 August 1971, 1, 52.

191. "The Apollo 11 Astronauts: Recognition of Religion?," *Christianity Today*, 18 July 1969, 31–32.

192. Hansen, *First Man*, 33–35.

193. Edgar Mitchell, *The Way of the Explorer: An Apollo Astronaut's Journey Through the Material and Mystical Worlds*, with Dwight Williams (New York: G. P. Putnam's Sons, 1996), 11, 18–19.

194. Walter Cunningham, *The All-American Boys* (New York: Macmillan, 1977), 26–27.

195. "Apollo 11 Astronauts."

196. Farmer and Hamblin, *First on the Moon*, 94; Wendt and Still, *Unbroken Chain*, 132.

197. Aldrin, "Communion in Space," 3.

198. Apollo 11 PAO Mission Commentary Transcript, 577, NASA/JSC History Portal, www.jsc.nasa.gov/history/mission_trans/apollo11.htm.

199. For an insightful and sympathetic discussion of Aldrin's personality, see Andrew Smith, *Moondust: In Search of the Men Who Fell to Earth* (London: Bloomsbury, 2005), 80–107.

200. Mailer, *Fire on the Moon*, 274.

201. Col. Edwin E. "Buzz" Aldrin Jr., *Return to Earth*, with Wayne Warga (New York: Random House, 1973), 99; Farmer and Hamblin, *First on the Moon*, 40–41, 307–8.

202. Farmer and Hamblin, *First on the Moon*, 40.

203. In his recent memoir, describing his life since *Apollo 11*, Aldrin notes: "Perhaps, if I had it to do over again, I would not choose to celebrate communion." It was, he observes, "a Christian sacrament, and we had come to the moon in the name of all mankind—be they Christians, Jews, Muslims, animists, agnostics, or atheists. But at the time I could think of no better way to acknowledge the enormity of the Apollo 11 experience than by giving thanks to God." Buzz Aldrin, *Magnificent Desolation: The Long Journey Home from the Moon*, with Ken Abraham (London: Bloomsbury, 2009), 27.

204. Farmer and Hamblin, *First on the Moon*, 267–68.

CHAPTER TWO: Signals of Transcendence

1. Søren Kierkegaard, *Concluding Unscientific Postscript to Philosophical Fragments*, translated by Howard V. Hong and Edna H. Hong (Princeton, NJ: Princeton University Press, 1992), 204.

2. Martin J. Heinecken, *God in the Space Age* (Philadelphia: Winston, 1959), 173.

3. See, e.g., Clifford J. Stevens, *Astrotheology: For the Cosmic Adventure* (Techny, IL: Divine Word, 1969).

4. "Towards a Hidden God," *Time*, 8 April 1966, www.time.com; Thomas J. J. Altizer and William Hamilton, *Radical Theology and the Death of God* (Harmondsworth, UK: Penguin, 1968).

5. For a still valuable discussion of the importance of height symbolism for Christianity and other religions, see Edwyn Bevan, *Symbolism and Belief* (London: George Allen & Unwin, 1938).

6. E. L. Mascall, *Christian Theology and Natural Science: Some Questions on their Relations* (London: Longmans, Green, 1956), 27.

7. Rudolf Bultmann, *New Testament and Mythology, and Other Basic Writings* (London: SCM, 1985), 1–2.

8. See Alfred North Whitehead, *Process and Reality*, edited by David Ray Griffin and Donald W. Sherburne, 2nd ed. (New York: Free Press, 1978); and Pierre Teilhard de Chardin, *The Phenomenon of Man*, translated by Bernard Wall (1975; reprint, New York: Perennial, 2002), originally published in French as *Le phénomène humain* (Paris: Editions du Seuil, 1955).

9. See Karl Barth, *Church Dogmatics*, vol. 3, *The Doctrine of Creation, Part One*, edited by G. W. Bromley and T. F. Torrance (Edinburgh: T & T Clark, 1958), 3–22.

10. See Paul Tillich, *Systematic Theology*, vol. 2, *Part III: Existence and the Christ* (London: SCM, 1978).

11. John A. T. Robinson, *Honest to God* (London: SCM, 1963).

12. Ibid., 11–18.

13. Altizer and Hamilton, *Radical Theology*, 39–40.

14. "Religionless Christianity," *Time*, 12 April 1963, www.time.com.

15. Jeffrey K. Hadden, *The Gathering Storm in the Churches* (Garden City, NY: Anchor Books, 1970), 20.

16. John A. T. Robinson, *Honest to God*, 15.

17. For details of the antireligious campaign, see Philip Walters, "A Survey of Soviet Religious Policy," in *Religious Policy in the Soviet Union*, edited by Sabrine Petra Ramet (Cambridge: Cambridge University Press, 1993), 20–23.

18. "Russian Writer Asserts Rockets Cast Doubt on Existence of God," *New York Times*, 23 January 1959, 3; "Saw Nothing in Space to Lead Him to Believe God Exists, Titov Says," *Washington Post*, 7 May 1962, A3.

19. C. L. Sulzberger, "Foreign Affairs: Paradise and Old Noah Khrushchev," *New York Times*, 9 September 1961, 18.

20. C. S. Lewis, *Christian Reflections*, edited by Walter Hooper (London: Geoffrey Bles, 1967), 168.

21. Jalmar Johnson, "Cherubims and Astronauts," *Christian Century*, 6 September 1961, 1456–57.

22. David H. C. Read, "Sputnik and the Angels," *Christianity Today*, 9 December 1957, 9–11.

23. Dan L. Thrapp, "Are We Really Alone? The Beginning of an Answer," *Los Angeles Times*, 13 July 1969, F1–F2.

24. Tim Zell, "Some Questions & Answers on the Church of All Worlds," *Green Egg*, 1 November 1972, 2. For a useful account of the Church of All Worlds, see Margot Adler, *Drawing Down the Moon: Witches, Druids, Goddess-Worshippers and Other Pagans in America* (London: Penguin, 2006), 300–334.

25. To identify God with the natural universe, Tillich had observed, was to deny "the infinite distance between the whole of finite things and their infinite ground." It was also to make the term *God* itself "semantically superfluous." Tillich, *Systematic Theology*, 2:7.

26. See, e.g., Adler, *Drawing Down the Moon*; and Robert Gottlieb, *Forcing the Spring: The Transformation of the American Environmental Movement* (Washington, DC: Island, 1993), 81–114.

27. "Apollo 8 Flight Journal," edited by W. David Woods and Frank O'Brien, "Day 4: Lunar Orbits 1, 2 and 3," NASA History Division, http://history.nasa.gov/ap08fj/index.htm.

28. McCurdy, *Space and the American Imagination*, 139–47.

29. Steven J. Dick, *Life on Other Worlds: The 20th-Century Extraterrestrial Life Debate* (Cambridge: Cambridge University Press, 1998), 43–65; Carl Sagan, *The Cosmic Connection: An Extraterrestrial Perspective* (London: Coronet, 1975), 81–93.

30. Robert A. Heinlein, *Stranger in a Strange Land* (1965; reprint, London: New English Library, 1978), first published in 1961.

31. Adler, *Drawing Down the Moon*, 305–26.

32. Archibald MacLeish, "A Reflection: Riders on Earth Together, Brothers in Eternal Cold," *New York Times*, 25 December 1968, 1.

33. *NBC Nightly News*, 22 April 1970, Vanderbilt Television News Archive, Nashville.

34. Adler, *Drawing Down the Moon*, 317–20; James Lovelock, *Gaia: A New Look at Life on Earth* (1979; reprint, Oxford: Oxford University Press, 1987). See also Poole, *Earthrise*, 170–78. The connection between the visual perspectives offered by the space age and descriptions of the earth as a complete living system was also evident in the various editions of the *Whole Earth Catalog*, jointly produced by Stewart Brand and the Portola Institute from 1968 to 1975. The catalog provided readers with information about how to access both the conceptual and the practical tools and skills required to build new communities based on ecological principles. The cover of each edition of the catalog featured a NASA photograph of the earth. See, e.g., *The (Updated) Last Whole Earth Catalog: Access to Tools* (San Francisco: Point, 1975).

35. Anne Morrow Lindbergh, "The Heron and the Astronaut," *Life*, 28 February 1969, 26.

36. Mark Oppenheimer, *Knocking on Heaven's Door: American Religion in the Age of Counterculture* (New Haven, CT: Yale University Press, 2003), 11–14.

37. Robert S. Ellwood Jr., "Church of All Worlds," *Green Egg*, 21 September 1973, 5–6.

38. Eliade, *Sacred and the Profane*; Philip Jenkins, *Dream Catchers: How Mainstream America Discovered Native Spirituality* (Oxford: Oxford University Press, 2004), 154–66.

39. Berger, *Rumour of Angels*, 62–71. See also Herbert W. Richardson and Donald R. Cutler, eds., *Transcendence* (Boston: Beacon, 1969).

40. Dean M. Kelley, *Why Conservative Churches Are Growing: A Study in the Sociology of Religion* (New York: Harper & Row, 1972); David F. Wells and John D. Woodbridge, eds., *The Evangelicals: What They Believe, Who They Are, Where They Are Changing* (Nashville: Abingdon, 1975).

41. In 1975, 52% of Americans asserted that their concept of God accorded with the following description: "The Supreme Being who created the earth and who rewards and punishes everyone in it." Roper Organization poll, 6–13 December 1975, Roper Center for Public Opinion Research, University of Connecticut, www.ropercenter.uconn.edu/. In 1968, 85% of respondents declared a belief in heaven. Gallup poll, 26 June–1 July 1968, Roper Center. In 1978, 56% claimed a belief in angels. Gallup poll, 24–27 February 1978, Roper Center. On hell and life after death, see Around and About, *Living Church*, 26 January 1969, 2; on devils, see Tidings, *Time*, 29 April 1974, www.time.com.

42. "Has God Forsaken the World?," *Christianity Today*, 19 December 1969, 20–21.

43. "Humanism and the Churches," ibid., 10 April 1970, 32–33.

44. "Does God Answer Prayer?," ibid., 8 May 1970, 27.

45. Billy Graham, quoted in "Praying the Space Heroes Home," ibid., 37.

46. David M. Jacobs, *The UFO Controversy in America* (Bloomington: Indiana University Press, 1975), 264–65. A poll in 1973 found that 54% of those who had heard or read about UFOs (94% of the sample) believed that they were "something real," and 11% of the same subset asserted that they had seen something they thought was a UFO. Gallup poll, 2–5 November 1973, Roper Center.

47. John Weldon, *UFOs: What on Earth is Happening? The Coming Invasion*, with Zola Levitt (New York: Bantam Books, 1976).
48. "Billy Graham Says Moon Trip Tribute to Creator," *Bucks County (PA) Courier Times*, 19 July 1969, 4, www.NewspaperArchive.com.
49. Billy Graham, *Angels: God's Secret Agents* (London: Hodder & Stoughton, 1975), 20–25.
50. Mascall, *Christian Theology and Natural Science*, 24–32. See also Karl Heim, *Christian Faith and Natural Science* (London: SCM, 1953).
51. See, e.g., Clifford Will, "The Renaissance of General Relativity," in *The New Physics*, edited by Paul Davies (Cambridge: Cambridge University Press, 1992), 7–33.
52. David M. Donahue, "Serving Students, Science, or Society? The Secondary School Physics Curriculum in the United States, 1930–65," *History of Education Quarterly* 33 (1993): 321–52.
53. Barry H. Downing, *The Bible and Flying Saucers* (London: Sphere Books, 1973), 130–33.
54. Ibid., 137, emphasis in original.
55. See esp. M. B. Foster, "The Christian Doctrine of Creation and the Rise of Modern Natural Science," *Mind* 43 (1934): 446–68; Foster, "Christian Theology and Modern Science of Nature (I)," ibid. 44 (1935): 439–66; Foster, "Christian Theology and Modern Science of Nature (II)," ibid. 45 (1936): 1–27; Robert K. Merton, "Science, Technology and Society in Seventeenth Century England," *Osiris* 4 (1938): 360–632; John Hedley Brooke, *Science and Religion: Some Historical Perspectives* (Cambridge: Cambridge University Press, 1991), 192–225; and Ronald L. Numbers, *The Creationists: From Scientific Creationism to Intelligent Design* (Cambridge, MA: Harvard University Press, 2006). For a more forthright disavowal of the conventional wisdom, see Rodney Stark, *For the Glory of God: How Monotheism Led to Reformations, Science, Witch-Hunts, and the End of Slavery* (Princeton, NJ: Princeton University Press, 2003), 121–97.
56. Jacques Ellul, *The Technological Society* (New York: Vintage, 1964), 415–23.
57. Lynn White Jr., "Dynamo and Virgin Reconsidered"; Noble, *Religion of Technology*, 12–15.
58. Noble, *Religion of Technology*, 110–14; Lisa McGirr, *Suburban Warriors: The Origins of the New American Right* (Princeton, NJ: Princeton University Press, 2001), 7–9.
59. James Gilbert, *Redeeming Culture: American Religion in an Age of Science* (Chicago: University of Chicago Press, 1997), 16.
60. See esp. Reinhold Niebuhr, *The Nature and Destiny of Man: A Christian Interpretation*, vol. 1, *Human Nature* (Louisville, KY: Westminster John Knox, 1996).
61. National Council of Churches, "Some Hopes and Concerns of the Church in the Nuclear-Space Age," 5 December 1957, in Martha Wheeler George, *Impact of Space on Religion*, 38.
62. See esp. William Hamilton, "The New Optimism—From Prufrock to Ringo," in Altizer and Hamilton, *Radical Theology*, 157–68.
63. See, e.g., Addison H. Leitch, "The View from the Moon," *Christianity Today*, 28 February 1969, 51; and "Our Foothold in the Heavens," ibid., 22 August 1969, 22–23.
64. "Pope Paul Hails Moon Explorers," *Bridgeport (CT) Post*, 21 July 1969, 9, www.NewspaperArchive.com.
65. Richard M. Nixon, "Remarks to Apollo 11 Astronauts Aboard the U.S.S. *Hornet* Following Completion of Their Lunar Mission," 24 July 1969, *Public Papers of the Presi-*

dents: Richard M. Nixon, 1969 (Washington, DC: Government Printing Office, 1971), 541–43; "Graham Disputes Nixon on 'Week,'" *Washington Post*, 26 July 1969, A10.

66. David Kucharsky, "The Lunar Landing," *Christianity Today*, 1 August 1969, 32; "Man's New Domain," ibid., 22 August 1969, 41–42.

67. Paul Tillich, "Man, the Earth and the Universe," *Christianity and Crisis*, 25 June 1962, 108–12.

68. Neufeld, *Von Braun*, 229–30.

69. Von Braun, handwritten notes for International Christian Leadership World Conference, 8 July 1965, folder: "Misc. office notes, memos, letter drafts, etc. 1963–68," box 44, von Braun Papers. International Christian Leadership drew a good portion of its early membership from the ranks of former Nazis and others who had served the Nazi state during the Second World War. See Sharlet, *Family*, 144–80.

70. Fallaci, *If the Sun Dies*, 206–32.

71. Mailer, *Fire on the Moon*, 61–65.

72. Ibid., 105.

73. Ibid., 313, 312, 76.

74. "New Priorities," *New York Times*, 21 July 1969, 16.

75. See, e.g., Richard M. Nixon, "Remarks to American Field Service Students," 22 July 1969, *Public Papers of the Presidents: Richard M. Nixon, 1969*, 533.

76. McQuaid, "Selling the Space Age."

77. "Good Earth?," *Newsweek*, 7 July 1969, 57.

78. R. Buckminster Fuller, *Operating Manual for Spaceship Earth* (Carbondale: Southern Illinois University Press, 1969).

79. "Good Earth?," 59.

80. *CBS Evening News*, 14 April 1970, Vanderbilt Television News Archive.

81. "Apollo 8 Flight Journal," ed. Woods and O'Brien, "Day 3: The Green Team," NASA History Division, http://history.nasa.gov/ap08fj/index.htm.

82. For valuable discussions of the Apollo *Earthrise* and whole-earth images, see Denis Cosgrove, "Contested Global Visions: *One-World*, *Whole-Earth*, and the Apollo Space Photographs," *Annals of the Association of American Geographers* 84 (1994): 270–94; and Poole, *Earthrise*.

83. Paul Boyer, *When Time Shall Be No More: Prophecy Belief in Modern American Culture* (Cambridge: Belknap Press of Harvard University Press, 1992), 10.

84. Hal Lindsey, *The Late Great Planet Earth*, with C. C. Carlson (Grand Rapids, MI: Zondervan, 1970). For a valuable discussion of Lindsey's writings, see Daniel Wojcik, *The End of the World As We Know It: Faith, Fatalism and Apocalypse in America* (New York: New York University Press, 1997), 37–59.

85. See esp. J. G. Ballard, "The Cage of Sand," in *Memories of the Space Age* (Sauk City, WI: Arkham House, 1988), 3–27; Michael Crichton, *The Andromeda Strain* (New York: Knopf, 1969); *Night of the Living Dead*, directed by George A. Romero (1968; London: Optimum Home Releasing, 2008), Blu-ray; *Planet of the Apes*, directed by Franklin J. Schaffner (1968; Los Angeles: Twentieth Century Fox, 2001), DVD; and *Silent Running*, directed by Douglas Trumbull (1972; Culver City, CA: UCA, 2008), DVD.

86. Weldon, *UFOs*; Robert W. Balch, "Waiting for the Ships: Disillusionment and the Revitalization of Faith in Bo and Peep's UFO Cult," in *The Gods Have Landed: New Religions from Other Worlds*, edited by James R. Lewis (Albany: State University of New York Press, 1995), 137–66; Wojcik, *End of the World*, 180–85.

87. McMillen, "Space Rapture," 85–86; Fallaci, *If the Sun Dies*, 14–15.

88. Paul R. Ehrlich, *The Population Bomb* (New York: Ballantine Books, 1968); Donella H. Meadows, Dennis L. Meadows, Jørgen Randers, and William W. Behrens III, *The Limits to Growth: A Report for the Club of Rome's Project on the Predicament of Mankind* (New York: Universe Books, 1972).

89. See esp. Gerard K. O'Neill, "The Colonization of Space," *Physics Today* 27 (September 1974): 32–40.

90. Bryce Harlow to Bob Haldeman, 14 April 1970, folder: "[Ex] RM 2 Prayers—Prayer Periods [1-69/12-70]," 2 of 2, box 3, Subject Files: Religious Matters, White House Central Files, Nixon Presidential Materials, National Archives, College Park, MD.

91. Richard M. Nixon, "Remarks at a Special Church Service in Honolulu," 19 April 1970, *Public Papers of the Presidents: Richard M. Nixon, 1970* (Washington, DC: Government Printing Office, 1971), 370–72.

92. Rev. R. D. Gillquist et al. to Richard M. Nixon, 28 April 1970, folder: "[Ex] OS 3 5/1/70–6/30/70," box 7, Subject Files: Outer Space, White House Central Files, Nixon Presidential Materials, National Archives, College Park.

93. Richard M. Nixon, "Statement About the Space Program," 19 December 1972, *Public Papers of the Presidents: Richard M. Nixon, 1972* (Washington, DC: Government Printing Office, 1974), 1157–59.

94. Lynn White Jr., "Historical Roots," 75–94.

95. See esp. Francis A. Schaeffer, *Pollution and the Death of Man: The Christian View of Ecology* (Wheaton, IL: Tyndale House, 1970). For a discussion of theological responses to the environmental crisis, see Michael S. Northcott, *The Environment and Christian Ethics* (Cambridge: Cambridge University Press, 1996), 124–63.

96. McQuaid, "Selling the Space Age"; W. Henry Lambright, "NASA and the Environment: Science in a Political Context," in Dick and Launius, *Societal Impact of Spaceflight*, 313–30. Roger D. Launius has attributed NASA's environmental turn in the 1970s to the influence of religious principles of stewardship, as imparted by James C. Fletcher, the agency's administrator, who was a "devout Mormon." Launius, "A Western Mormon in Washington, D.C.: James C. Fletcher, NASA, and the Final Frontier," *Pacific Historical Review* 64 (1995): 217–41. There is evidence, however, that Fletcher was personally skeptical about the merits of the "environmental theme." McQuaid, "Selling the Space Age," 128–29; John Donnelly to Fletcher, "Environmental Theme," 3 July 1973, folder 5: "Correspondence (25 April–17 July 1973)," box 16, Fletcher Papers.

97. Erich von Daniken, *Chariots of the Gods?* (London: Corgi, 1971); von Daniken, *Return to the Stars* (London: Corgi, 1972). In the years 1970–76 more than 12 million copies of von Daniken's works were sold in the United States. Ronald Story, *The Space Gods Revealed: A Close Look at the Theories of Erich von Daniken* (London: New English Library, 1976), 17. The notion that mankind may have derived its knowledge and skills from an early alien intervention was not unique to von Daniken. See, e.g., *2001: A Space Odyssey*, directed by Stanley Kubrick (1968; Burbank, CA: Warner Home Video, 2006), DVD.

98. C. G. Jung, *Flying Saucers: A Modern Myth of Things Seen in the Skies* (London: Routledge & Kegan Paul, 1959).

99. Friedrich Nietzsche, *The Gay Science* (Cambridge: Cambridge University Press, 2001), 120.

100. Dick, *Life on Other Worlds*, 78–88,

101. Harlow Shapley, *Of Stars and Men: The Human Response to an Expanding Universe* (London: Elek Books, 1958), 67–70; Walter Sullivan, *We Are Not Alone: The Search for Intelligent Life on Other Worlds* (Harmondsworth, UK: Penguin, 1970), 64–69.

102. Dick, *Life on Other Worlds*, 23–24, 53–65, 188–92. See also Steven J. Dick and James E. Strick, *The Living Universe: NASA and the Development of Astrobiology* (New Brunswick, NJ: Rutgers University Press, 2004).

103. Steven J. Dick, *Plurality of Worlds: The Origins of the Extraterrestrial Life Debate from Democritus to Kant* (Cambridge: Cambridge University Press, 1982).

104. Michael J. Crowe, *The Extraterrestrial Life Debate, 1750–1900: The Idea of a Plurality of Worlds from Kant to Lowell* (Cambridge: Cambridge University Press, 1986). See also David Cressy, "Early Modern Space Travel and the English Man in the Moon," *American Historical Review* 111 (2006): 961–82.

105. Thomas Paine, *The Age of Reason* (Whitefish, MT: Kessinger, 2004), 49–59.

106. William Whewell, *The Plurality of Worlds* (Boston: Gould & Lincoln, 1854).

107. William Derham, *Astro-Theology; Or, a Demonstration of the Being and Attributes of GOD* (London: W. & J. Innys, 1721), xlii.

108. Crowe, *Extraterrestrial Life Debate*, 549–50.

109. Erich Robert Paul, *Science, Religion, and Mormon Cosmology* (Urbana: University of Illinois Press, 1992).

110. Father Domenico Grasso, quoted in Wolfgang D. Müller, *Man Among the Stars* (London: George G. Harrap, 1958), 222–23.

111. C. S. Lewis, *Christian Reflections*, 176.

112. C. S. Lewis, *The World's Last Night, and Other Essays* (New York: Harcourt, 2002), 83–92. In addition, see Lewis's two novels about space travel, *Out of the Silent Planet* (New York: Scribner, 1996) and *Perelandra* (New York: Scribner, 1996).

113. Donald N. Michael, "Proposed Studies on the Implications of Peaceful Space Activities for Human Affairs: A Report Prepared for the Committee on Long-Range Studies of the National Aeronautics and Space Administration by the Brookings Institution," December 1960, www.nicap.org/brookingsdir.htm.

114. See, e.g., Heinecken, *God in the Space Age*, 122; and Charles K. Robinson, "The Space Age and Christology," *Theology Today* 14 (January 1963): 500–509.

115. Nietzsche, *Gay Science*, 119–20.

116. George Dugan, "Science is Called No Peril to Faith," *New York Times*, 20 June 1959, 12.

117. Arthur C. Clarke, *Profiles of the Future: An Inquiry into the Limits of the Possible* (London: Indigo, 2000), 89.

118. For two exceptions, see the statement of Father Agostino Gemilli in "Space Theology," *Time*, 19 September 1955, www.time.com; and Joseph A. Breig, "Man Stands Alone," *America*, 26 November 1960, 294–95.

119. Francis J. Heyden, "The Higher Promise of Space Exploration," in *Space: Its Impact on Man and Society*, edited by Lillian Levy (New York: Norton, 1965), 180–84.

120. Norman Lamm, "The Religious Implications of Extraterrestrial Life," in *Challenge: Torah Views on Science and its Problems*, edited by Aryeh Carmell and Cyril Domb (Jerusalem: Association of Orthodox Jewish Scientists, 1978), 392.

121. Tillich, "Man, the Earth, and the Universe," 110. See also Tillich, *Systematic Theology*, 2:95.

122. For speculations on the principal possibilities, see "The Theology of Saucers,"

Time, 18 August 1952, www.time.com; Daniel C. Raible, "Rational Life in Outer Space," *America*, 13 August 1960, 532–35; T. J. Zubek, "Theological Questions on Space Creatures," *American Ecclesiastical Review* 145 (1961): 393–99; and C. S. Lewis, *The World's Last Night*, 83–92.

123. C. S. Lewis, *The World's Last Night*, 86.

124. For such a view, see E. L. Milne, *Modern Cosmology and the Christian Idea of God* (Oxford: Clarendon, 1952), 153–54.

125. Edward U. Condon et al., *Scientific Study of Unidentified Flying Objects* (Boulder: Regents of the University of Colorado, 1968), 38, www.ncas.org/condon/index.html.

126. Mascall, *Christian Theology and Natural Science*, 37–39; Tillich, *Systematic Theology*, 2:95–96.

127. See, e.g., Tillich, *Systematic Theology*, 2:96; and Heinecken, *God in the Space Age*, 142–45.

128. W. Norman Pittenger, *The Word Incarnate: A Study of the Doctrine of the Person of Christ* (Welwyn, UK: James Nisbet, 1959), 248–51. See also Pittenger, "Christianity and the Man on Mars," *Christian Century*, 20 June 1956, 747–48.

129. Charles K. Robinson, "Space Age and Christology," 506–8. See also L. C. McHugh, "Others Out Yonder," *America*, 26 November 1960, 295–97.

130. Mascall, *Christian Theology and Natural Science*, 45.

131. Dick, *Life on Other Worlds*, 254–60.

132. See esp. John D. Barrow and Frank J. Tipler, *The Anthropic Cosmological Principle* (Oxford: Oxford University Press, 1986).

133. See, e.g., Michael A. Corey, *The God Hypothesis: Discovering Design in Our "Just Right" Goldilocks Universe* (Lanham, MD: Rowman & Littlefield, 2001).

134. Rodney W. Johnson, "Spiritual Implications of Exploring the Moon," *Christianity Today*, 7 January 1972, 4–6.

135. See Turner, *Bill Bright and Campus Crusade for Christ*, 119–46; Robert S. Ellwood Jr., *One Way: The Jesus Movement and Its Meaning* (Englewood Cliffs, NJ: Prentice-Hall, 1973); Stephen Prothero, *American Jesus: How the Son of God Became a National Icon* (New York: Farrar, Straus & Giroux, 2003), 124–57; and Richard Wightman Fox, *Jesus in America: Personal Savior, Cultural Hero, National Obsession* (New York: HarperOne, 2004), 376–80.

136. "Dabbling in Exotheology," *Time*, 24 April 1978, www.time.com.

137. Jack A. Jennings, "UFOs: The Next Theological Challenge?," *Christian Century*, 22 February 1978, 184–89.

138. Ted Peters, "Exo-Theology: Speculations on Extraterrestrial Life," in James R. Lewis, *Gods Have Landed*, 188; Steven J. Dick, "Cosmotheology: Theological Implications of the New Universe," in *Many Worlds: The New Universe, Extraterrestrial Life and the Theological Implications*, edited by Steven J. Dick (Philadelphia: Templeton Foundation Press, 2000), 191–210.

139. Mailer, *Fire on the Moon*, 29.

CHAPTER THREE: Into the Other World

1. Glenn, *John Glenn*, 263, 276. Glenn's line about four sunsets is missing from the official transcript of his recorded responses contained in John Glenn, "Brief Summary of MA-6 Orbital Flight," 20 February 1962, enclosed in Robert B. Voas, "MA-6 Pilot's

Debriefing," 22 February 1962, in *Exploring the Unknown: Selected Documents from the History of the U.S. Civil Space Program*, vol. 7, *Human Spaceflight: Projects Mercury, Gemini, and Apollo*, edited by John M. Logsdon with Roger D. Launius (Washington, DC: NASA History Division, 2008), 223–28.

2. Glenn, *John Glenn*, 165–75.
3. Ibid., 5. See also Carpenter et al., *We Seven*, 12–15; and Wolfe, *Right Stuff*, 116–19.
4. James L. Kauffman, *Selling Outer Space: Kennedy, the Media, and Funding for Project Apollo, 1961–1963* (Tuscaloosa: University of Alabama Press, 1994), 43.
5. Carpenter et al., *We Seven*, 409–10. On "real-time" public awareness of the heat-shield problem, see "The Flight," *Time*, 2 March 1962, www.time.com; and Michael Allen, *Live From the Moon: Film, Television and the Space Race* (London: I. B. Tauris, 2009), 89–90.
6. Glenn, *John Glenn*, 278–83.
7. John W. Finney, "Astronauts Give View," *New York Times*, 1 March 1962, 1, 15.
8. *Orbital Flight of John H. Glenn, Jr.: Hearing Before the Senate Committee on Aeronautical and Space Sciences*, 87th Cong. 1, 13, 16 (1962).
9. "John H. Glenn: An Astronaut and His Faith," *Christianity Today*, 16 March 1962, 31.
10. Carpenter et al., *We Seven*, 397.
11. John H. Glenn, "Faith is a Star," *Washington Evening Star*, 7 December 1963.
12. "Orbital Flight of John H. Glenn, Jr.," 13.
13. William G. McLoughlin, *Revivals, Awakenings, and Reform: An Essay on Religion and Social Change in America, 1607–1977* (Chicago: University of Chicago Press, 1978), 179–216; Ellwood, *Sixties Spiritual Awakening*.
14. Mayo Mohs, "God, Man and Apollo," *Time*, 1 January 1973, www.time.com. See also Paul Recer, "The Astros—Men Changed in Heaven," *Los Angeles Herald-Examiner*, 30 July 1972, 5–8; "Space: The Greening of the Astronauts," *Time*, 11 December 1972, www.time.com; and George W. Cornell, "Astronauts Find God in Space," *San Diego Union*, 19 May 1973.
15. C. S. Lewis, *Christian Reflections*, 171.
16. Catherine L. Albanese, *A Republic of Mind and Spirit: A Cultural History of American Metaphysical Religion* (New Haven, CT: Yale University Press, 2007), 1–7.
17. Ann Taves, *Fits, Trances, and Visions: Experiencing Religion and Explaining Experience from Wesley to James* (Princeton, NJ: Princeton University Press, 1999).
18. Friedrich Schleiermacher, *The Christian Faith* (London: T & T Clark, 1999), 12, 27. For a useful discussion of Kant and Schleiermacher, see Wayne Proudfoot, *Religious Experience* (Berkeley: University of California Press, 1985), 1–23.
19. Rudolf Otto, *The Idea of the Holy* (London: Oxford University Press, 1926), 5–11.
20. Richard M. Bucke, *Cosmic Consciousness: A Study in the Evolution of the Human Mind* (Philadelphia: Innes & Sins, 1905), 8.
21. William James, *The Varieties of Religious Experience* (New York: Barnes & Noble Classics, 2004), 19.
22. For such assumptions, see James H. Leuba, *The Psychology of Religious Mysticism* (London: Kegan Paul, Trench, Trubner, 1929); and Edwin D. Starbuck, *The Psychology of Religion: An Empirical Study of the Growth of Religious Consciousness* (New York: Walter Scott, 1914).
23. In October 1954 the most prominent of these preachers, Billy Graham, appeared

on the cover of *Time*. The accompanying article describes in some detail the process by which the revivals achieved new "converts." See "The New Evangelist," *Time*, 25 October 1954, www.time.com.

24. Gallup poll, 15 April 1962, in Gallup, *Gallup Poll*, vol. 3, *1959–1971*, 1762–63.

25. Stark and Glock, *American Piety*, 129–34.

26. See, e.g., Tillich, *Systematic Theology*, vol. 3, esp. "Part IV: Life and the Spirit"; C. G. Jung, *The Archetypes of the Collective Unconscious*, 2nd ed. (London: Routledge, 1991); and Eliade, *Myth of the Eternal Return*.

27. See esp. Abraham H. Maslow, *Motivation and Personality* (New York: Harper's, 1954); and Maslow, *Religions, Values, and Peak Experiences* (New York: Penguin Compass, 1976).

28. "Christianity Today–Gallup Poll," 14.

29. Richard Kyle, *The New Age Movement in American Culture* (Lanham, MD: University Press of America, 1995), 15–16.

30. Robert Wuthnow, "The New Religions in Social Context," in *The New Religious Consciousness*, edited by Charles Y. Glock and Robert N. Bellah (Berkeley: University of California Press, 1976), 269–74.

31. John F. Kennedy, "Remarks Following the Orbital Flight of Col. John H. Glenn, Jr.," 20 February 1962, *Public Papers of the Presidents: John F. Kennedy, 1962*, 150.

32. Byrnes, *Politics and Space*, 50–52; Kauffman, *Selling Outer Space*, 32–45.

33. Wayne Franklin, *Discoverers, Explorers, Settlers: The Diligent Writers of Early America* (Chicago: University of Chicago Press, 1979), 22.

34. William D. Phillips Jr. and Carla Rahn Phillips, *The Worlds of Christopher Columbus* (Cambridge: Cambridge University Press, 1992), 220; John Leddy Phelan, *The Millennial Kingdom of the Franciscans in the New World* (Berkeley: University of California Press, 1970), 19–23.

35. Christopher Columbus, Letter on the Fourth Voyage (1503), Doc. AJ-068, American Journeys Collection, Wisconsin Historical Society Digital Library and Archives, http://content.wisconsinhistory.org/u?/aj,4444.

36. Robert Lawson-Peebles, *Landscape and Written Expression in Revolutionary America: The World Turned Upside Down* (Cambridge: Cambridge University Press, 1988), 196–222.

37. Immanuel Kant, *Critique of Practical Reason*, translated by T. K. Abbot (1879; reprint, Amherst, NY: Prometheus Books, 1996), 191, originally published in German as *Kritik der praktischen Vernunft* (Riga: Hartknoch, 1788); Marjorie Hope Nicolson, *Mountain Gloom and Mountain Glory: The Development of the Aesthetics of the Infinite* (Ithaca, NY: Cornell University Press, 1959).

38. See esp. Immanuel Kant, *The Critique of Judgement*, translated by James Creed Meredith (Oxford: Oxford University Press, 1952), originally published in German as *Kritik der Urteilskraft* (Berlin and Libau: bey Lagarde & Friederich, 1790).

39. Franklin, *Discovers, Explorers, Settlers*, 24–33; Barbara Novak, *Nature and Culture: American Landscape and Painting, 1825–1875* (Oxford: Oxford University Press, 2007), 3–14; Martha A. Sandweiss, *Print the Legend: Photography and the American West* (New Haven, CT: Yale University Press, 2002), 204–6.

40. Aaron Sachs, *The Humboldt Current: A European Explorer and His American Disciples* (Oxford: Oxford University Press, 2007), 74–76.

41. For Humboldt's influence, see ibid.; and William H. Goetzmann, *New Lands,*

New Men: America and the Second Great Age of Discovery (New York: Viking, 1986), 150–93.

42. Henry David Thoreau, *Walden*, in *The Portable Thoreau*, edited by Carl Bode (Harmondsworth, UK: Penguin, 1977), 559.

43. See esp. Catherine L. Albanese, *Nature Religion in America: From the Algonkian Indians to the New Age* (Chicago: University of Chicago Press, 1990).

44. Leo Marx, *The Machine in the Garden: Technology and the Pastoral Ideal in America* (Oxford: Oxford University Press, 2000), 11–15.

45. On that tradition, see Henry Nash Smith, *Virgin Land: The American West as Symbol and Myth* (Cambridge, MA: Harvard University Press, 1950).

46. Marx, *Machine in the Garden*, 373–74.

47. Walt Whitman, "To a Locomotive in Winter," in *The Complete Poems* (London: Penguin, 2004), 482.

48. Henry Adams, *The Education of Henry Adams* (New York: Oxford University Press, 1999), 318–19.

49. Margret Dreikhausen, *Aerial Perception: The Earth as Seen from Aircraft and Spacecraft and its Influence on Contemporary Art* (Philadelphia: Art Alliance, 1985), 30–22.

50. Charles A. Lindbergh, *Spirit of St. Louis*, 301–2.

51. Brant Clark and Ashton Graybiel, "The Break-Off Phenomenon," *Journal of Aviation Medicine* 28 (1957): 121–26.

52. David A. Simons, *Man High*, with Don A. Schanche (London: Sidgwick & Jackson, 1960), 175.

53. John Gillespie Magee, "High Flight," in *On the Wing: American Poems of Air and Space Flight*, edited by Karen Yelena Olsen (Iowa City: University of Iowa Press, 2005), 46.

54. On the poem's popularity among aviators, see Collins, *Carrying the Fire*, 199.

55. Charles A. Lindbergh, *Of Flight and Life*, 49–50.

56. Wohl, *Passion for Wings*, 138–43.

57. David A. Mindell, *Digital Apollo: Human and Machine in Spaceflight* (Cambridge, MA: MIT Press, 2008), 17–41.

58. Borman, *Countdown*, 114–15.

59. Charles A. Lindbergh, *Spirit of St. Louis*, 11–15.

60. Ibid., 389.

61. Charles A. Lindbergh, *Autobiography of Values*, 395.

62. Ibid., 12.

63. Franz Kafka, "The Aeroplanes at Brescia," in *The Transformation ('Metamorphosis'), and Other Stories* (London: Penguin, 1992), 1–10.

64. Wohl, *Passion for Wings*, 203–50.

65. Wohl, *Spectacle of Flight*, 25–27.

66. Fitzhugh Green, "What the World Thought of Lindbergh," 186.

67. John W. Ward, "The Meaning of Lindbergh's Flight," *American Quarterly* 10 (1958): 3–16.

68. Wohl, *Spectacle of Flight*, 308; Corn, *Winged Gospel*, 65–66.

69. See esp. Charles A. Lindbergh, *Of Flight and Life*, 51–52.

70. Charles A. Lindbergh, *Autobiography of Values*, 36–37, 304–5, 379–402.

71. Schiff, *Saint-Exupéry*, 53.

72. Saint-Exupéry, *Wind, Sand and Stars*, 102.

194 Notes to Pages 87–91

73. Sigmund Freud, *Leonardo da Vinci and a Memory of His Childhood* (New York: Norton, 1964).

74. Douglas C. Bond, *The Love and Fear of Flying* (New York: International Universities Press, 1952), 31.

75. Corn, *Winged Gospel*, 138.

76. Richard M. Nixon, "Proclamation 3919, National Day of Participation Honoring the Apollo 11 Mission," 16 July 1969, *Public Papers of the Presidents: Richard M. Nixon, 1969*, 518–19.

77. Wolfe, *Right Stuff*, 109–11.

78. By the end of the sixties, a recent expansion of academic religious studies programs as well as Carlos Castaneda's popular description, in *The Teachings of Don Juan*, of his own initiation into shamanic belief and ritual had carried the concept of the shaman into the field of public knowledge. See Eliade, *Myths, Dreams, and Mysteries*, 60–61; and Carlos Castaneda, *The Teachings of Don Juan: A Yaqui Way of Knowledge* (Berkeley: University of California Press, 1968).

79. Columbia Broadcasting System, *10:56:20PM 7/20/69: The Historic Conquest of the Moon as Reported to the American People by CBS News over the CBS Television Network* (New York, 1970), 13.

80. "Challenge in the Heavens," *Time*, 24 January 1969, www.time.com.

81. Nietzsche, *Gay Science*, 120.

82. Tillich, "Man, the Earth and the Universe."

83. J. Gordon Melton, "The Contactees: A Survey," in James R. Lewis, *Gods Have Landed*, 7–8.

84. See, e.g., "Go!," *Washington Post*, 21 February 1962, A24; Edward B. Lindaman, *Space: A New Dimension for Mankind* (New York: Harper & Row, 1969), 21–22; and the comment by Howard K. Smith on *ABC Evening News*, 21 July 1969, Vanderbilt Television News Archive. Almost without exception, such assertions were accompanied by reference to a passage in Samuel Eliot Morison and Henry Steele Commager's classic popular history text, *The Growth of the American Republic*, first published in 1930. See Samuel Eliot Morison, Henry Steele Commager, and William E. Leuchtenberg, *The Growth of the American Republic*, 7th ed. (Oxford: Oxford University Press, 1980), 15.

85. On the cultural history of the moon, see Bernd Brunner, *Moon: A Brief History* (New Haven, CT: Yale University Press, 2010); Paul Katzeff, *Moon Madness* (London: Robert Hale, 1990); Scott L. Montgomery, *The Moon and the Western Imagination* (Tucson: University of Arizona Press, 1999); and Rick Stroud, *The Book of the Moon* (London: Doubleday, 2009).

86. For speculations of this sort, see Charles Kuralt's report on the *CBS Evening News*, 23 December 1968, Vanderbilt Television News Archive; Paul O'Neill, "So Long to the Good Old Moon," *Life*, 4 July 1969, 47D–48D; and Mailer, *Fire on the Moon*, 14–15, 68, 228–36. According to Tom Stoppard, the initial inspiration for his 1972 play *Jumpers* was "my private thought that if and when men landed on the moon, something interesting would occur in the human psyche, that landing on the moon would be an act of destruction. There is a quotation when the first landing occurred from the Union of Persian Storytellers—if you can imagine such a thing—they claimed that it was somehow damaging of the livelihood of the storytellers. I understood that completely." "Tom Stoppard interview," by Brian Appleyard, *Sunday Times*, 8 June 2003,

www.bryanappleyard.com/article.php?page=9&article_id=10; see also Tom Stoppard, *Jumpers* (London: Faber & Faber, 1986).

87. Michael F. Robinson, *The Coldest Crucible: Arctic Exploration and American Culture* (Chicago: University of Chicago Press, 2006), 9, 160–64.

88. De Witt Douglas Kilgore, *Astrofuturism: Science, Race, and Visions of Utopia in Space* (Philadelphia: University of Pennsylvania Press, 2003).

89. Ballard, *Memories of the Space Age*, 90.

90. James H. Capshew, *Psychologists on the March: Science, Practice, and Professional Identity in America, 1929–1969* (Cambridge: Cambridge University Press, 1999), 143–47.

91. See Bernard E. Flaherty, ed., *Psychophysiological Aspects of Space Flight* (New York: Columbia University Press, 1961); Lang, *From Hiroshima to the Moon*, 507–28; and S. B. Sells and Charles A. Berry, "Human Requirements for Space Travel," in *Human Factors in Jet and Space Travel: A Medical-Psychological Analysis*, edited by Sells and Berry (New York: Ronald, 1961), 166–86.

92. Clark and Graybiel, "Break-Off Phenomenon"; Simons, *Man High*, 87–88, 176; Lang, *From Hiroshima to the Moon*, 524–25.

93. Philip Solomon, Herbert Leiderman, Jack Mendelson, and Donald Wexler, "Sensory Deprivation: A Review," *American Journal of Psychiatry* 114 (1957): 357–63.

94. Charles A. Lindbergh, *Spirit of St. Louis*, 389–91.

95. D. Ewen Cameron, Leonard Levy, Thomas Ban, and Leonard Rubenstein, "Sensory Deprivation: Effects upon the Functioning Human in Space Systems," in Flaherty, *Psychophysiological Aspects of Space Flight*, 225–37; Lang, *From Hiroshima to the Moon*, 522–24; John C. Lilly and Jay T. Shurley, "Experiments in Solitude, in Maximum Achievable Physical Isolation with Water Suspension, of Intact Healthy Persons," in Flaherty, *Psychophysiological Aspects of Space Flight*, 238–47; Jay T. Shurley, "Mental Imagery in Profound Experimental Sensory Isolation," in *Hallucinations*, edited by Louis J. West (New York: Grune & Stratton, 1962), 153–57; Shurley, "Profound Experimental Sensory Isolation," *American Journal of Psychiatry* 117 (1960): 539–45; Simons, *Man High*, 256–57; George R. Steinkamp and George T. Hauty, "Simulated Space Flights," in Flaherty, *Psychophysiological Aspects of Space Flight*, 75–79.

96. Philip Solomon, "Motivations and Emotional Reactions in Early Space Flights," in Flaherty, *Psychophysiological Aspects of Space Flight*, 272–77.

97. George E. Ruff and Edwin Z. Levy, "Psychiatric Evaluation of Candidates for Space Flight," *American Journal of Psychiatry* 116 (1959): 385–91; Patricia A. Santy, *Choosing the Right Stuff: The Psychological Selection of Astronauts and Cosmonauts* (Westport, CT: Praeger, 1994), 10–21. For a valuable account of the selection process from the candidates' perspective, see Carpenter and Stoever, *For Spacious Skies*, 163–95.

98. Charles L. Wilson, ed., "Project Mercury Candidate Evaluation Program," WADC Technical Report 59–505, December 1959, in Logsdon, *Exploring the Unknown*, 7:151–58.

99. For accounts of the Mercury astronauts' training regime, see Robert B. Voas, "Project Mercury Astronaut Training Program," 30 May 1960, in Logsdon, *Exploring the Unknown*, 7:161–72; Carpenter et al., *We Seven*, 170–207; and Loyd S. Swenson Jr., James M. Grimwood, and Charles C. Alexander, *The New Ocean: A History of Project Mercury* (Washington, DC: NASA, 1966), 235–48, 344–45, 413–19.

100. Carpenter et al., *We Seven*, 234, 284.

101. Ibid., 263–64, 436–37; Carpenter and Stoever, *For Spacious Skies*, 298–99.

102. Swenson, Grimwood, and Alexander, *New Ocean*, 213–18; Carpenter et al., *We Seven*, 251–52, 348–67, 389–92; Wolfe, *Right Stuff*, 307–8.

103. Carpenter et al. *We Seven*, 73. See also Wolfe, *Right Stuff*, 89.

104. On this point, see Christopher C. Kraft Jr., "A Review of Knowledge Acquired from the First Manned Satellite Program," [c. 1963], in Logsdon, *Exploring the Unknown*, 7:245–52.

105. Carpenter and Stoever, *For Spacious Skies*, 299; Santy, *Choosing the Right Stuff*, 28–31.

106. Fallaci, *If the Sun Dies*, 149.

107. Ibid., 292–93.

108. Carpenter et al., *We Seven*, 445–65; Carpenter and Stoever, *For Spacious Skies*, 243–95; M. Scott Carpenter, oral history interview by Michelle Kelly, 30 March 1998, NASA/JSC Oral History Project, 13, www.jsc.nasa.gov/history/oral_histories/oral_histories.htm.

109. For a critical account of Carpenter's mission, see Kraft, *Flight*, 163–70.

110. Wolfe, *Right Stuff*, 313.

CHAPTER FOUR: Perhaps a Meaning to Us

1. Carpenter et al., *We Seven*, 397, 451–52.

2. Alan Bean, *Apollo: An Eyewitness Account*, with Andrew Chaikin (Shelton, CT: Greenwich Workshop, 1998), 20; NASA, "Apollo 8 Technical Debriefing," 2 January 1969, 33–34, www.ibiblio.org/apollo/Documents/Apollo8-TechnicalDebriefing-Martin-1.pdf.

3. Bean, *Apollo*, 19–20; Andrew Chaikin, *Voices From the Moon: Apollo Astronauts Describe Their Lunar Experiences*, with Victoria Kohl (London: Viking, 2009), 36.

4. Farmer and Hamblin, *First on the Moon*, 202; William Anders, "Translunar Photos," Apollo 8 Flight Crew Log, box 2, Apollo Crew Logs, 1967–72, Flight Operations Directorate, Records of Johnson Space Center, National Archives Southwest Region, Fort Worth; Poole, *Earthrise*, 22–23.

5. Conrad and Klausner, *Rocket Man*, 150; Eugene Cernan and Don Davis, *The Last Man on the Moon: Astronaut Eugene Cernan and America's Race in Space* (New York: St. Martin's Griffin, 1999), 137; James B. Irwin, *To Rule the Night*, with William A. Emerson Jr. (Old Tappan, NJ: Spire Books, 1975), 43; Charlie Duke and Dotty Duke, *Moonwalker* (Nashville: Oliver Nelson, 1990), 175; Poole, *Earthrise*, 98–100.

6. Russell Schweickart, "No Frames, No Boundaries," in *Earth's Answer: Explorations of Planetary Culture at the Lindisfarne Conferences*, edited by Michael Katz, William P. Marsh, and Gail Gordon Thompson (New York: Harper & Row, 1977), 3–13.

7. Collins, *Carrying the Fire*, 245.

8. Mitchell, *Way of the Explorer*, 53.

9. David Scott and Alexei Leonov, *Two Sides of the Moon* (London: Simon & Schuster, 2004), 316–17; James B. Irwin, *To Rule the Night*, 72.

10. Edgar D. Mitchell, oral history interview by Sharee Scarborough, 3 September 1997, NASA/JSC Oral History Project, 31, 39–40, www.jsc.nasa.gov/history/oral_histories/oral_histories.htm.

11. Cunningham, *All-American Boys*, 133.

12. Borman, *Countdown*, 200.

13. James B. Irwin, *To Rule the Night*, 29.
14. Thomas K. Mattingley II, oral history interview by Rebecca Wright, 6 November 2001, NASA/JSC Oral History Project, 91–92, www.jsc.nasa.gov/history/oral_histories/oral_histories.htm.
15. Ibid., 92; Collins, *Carrying the Fire*, 444; Duke and Duke, *Moonwalker*, 227.
16. James B. Irwin, *To Rule the Night*, 29.
17. Carpenter et al., *We Seven*, 255, 290; Ray E. Boomhower, *Gus Grissom: The Lost Astronaut* (Indianapolis: Indiana Historical Society Press, 2004), 193.
18. "Appendix: MA-7 Air-Ground Voice Communications," in *Results of the Second U.S. Manned Orbital Space Flight, 24 May 1962*, NASA SP-6 (Washington, DC: NASA, 1962), http://history.nasa.gov/SP-6/contents.htm.
19. Borman, *Countdown*, 141. See also Cernan and Davis, *Last Man on the Moon*, 118; Cunningham, *All-American Boys*, 124; and Henry S. F. Cooper Jr., *A House in Space* (London: Granada, 1978), 121–33.
20. *Composite Air-to-Ground and Onboard Voice Tape Transcription of the GT-4 Mission* (Houston: NASA Manned Space Center, 1965), 56, http://insideksc.com/docs/transcripts/PDF/GT04_TEC.PDF
21. Schweickart, "No Frames, No Boundaries," 12.
22. Ibid., 11, 13.
23. Collins, *Carrying the Fire*, 385.
24. Borman, *Countdown*, 212.
25. Ibid., 224–25. The phrase "eternal cold" came from Archibald MacLeish, quoted by Borman in his address to Congress after the return of *Apollo 8*. "Joint Meeting of the Two Houses of Congress to Receive the Apollo 8 Astronauts," Cong. Rec. 368 (1969).
26. Collins, *Carrying the Fire*, 398.
27. Poole, *Earthrise*, 99.
28. Cernan and Davis, *Last Man on the Moon*, 208–9.
29. Frank White, *The Overview Effect: Space Exploration and Human Evolution* (Boston: Houghton Mifflin, 1987), 207.
30. Robert Zimmerman, *Genesis: The Story of Apollo 8* (New York: Dell, 1998), 261–62, 293–94.
31. Alfred M. Worden, *Hello Earth: Greetings from Endeavour* (Los Angeles: NASH, 1974), 51.
32. Mattingley, oral history interview, 6 November 2001, 89–90; Chaikin, *Voices from the Moon*, 122.
33. Duke and Duke, *Moonwalker*, 217.
34. Mitchell, *Way of the Explorer*, 55.
35. Ibid., 3.
36. Ibid., 17–18; Mitchell, oral history interview, 3 September 1997, 31–32.
37. Mitchell, *Way of the Explorer*, 3–4.
38. Ibid., 59.
39. Ibid., 3, emphasis in original.
40. See Farmer and Hamblin, *First on the Moon*, 201–3, 324.
41. Collins, *Carrying the Fire*, 393.
42. Farmer and Hamblin, *First on the Moon*, 324.
43. "Apollo 8 Flight Journal: Day 4: Lunar Orbits 1, 2 and 3."
44. Collins, *Carrying the Fire*, 397.

45. Scott and Leonov, *Two Sides of the Moon*, 289.
46. Duke and Duke, *Moonwalker*, 121.
47. "Apollo 11 Lunar Surface Journal: One Small Step," edited by Eric M. Jones, NASA History Division, www.hq.nasa.gov/alsj/a11/a11.html.
48. Ibid.
49. Chaikin, *Man on the Moon*, 187–88; Donald A. Beattie, *Taking Science to the Moon: Lunar Experiments and the Apollo Program* (Baltimore: Johns Hopkins University Press, 2001), 217–19.
50. Scott and Leonov, *Two Sides of the Moon*, 295.
51. James B. Irwin, *To Rule the Night*, 112–13.
52. Ibid., 4; "James Irwin—Testimony," undated, folder 12, container 24, Records of the International Christian Broadcasters, Collection 86, Billy Graham Center Archives, Wheaton, IL; Mary Irwin, *The Moon Is Not Enough*, with Magdalene Harris (London: Pickering & Inglis, 1979), 61.
53. On the Irwins' marital difficulties, see James B. Irwin, *To Rule the Night*, 164–66, 178–80; and Mary Irwin, *Moon Is Not Enough*, 63–101. On Irwin's discomfort during the voyage of *Apollo 15*, see James B. Irwin, *To Rule the Night*, 20–29.
54. James B. Irwin, *To Rule the Night*, 36–38.
55. Ibid., 43, 46.
56. Ibid., 52–53.
57. Scott and Leonov, *Two Sides of the Moon*, 308.
58. "Apollo 15 Lunar Surface Journal: The Genesis Rock," edited by Eric M. Jones, NASA History Division, http://history.nasa.gov/alsj/a15/a15j.html.
59. James B. Irwin, *To Rule the Night*, 57; "James Irwin—Testimony."
60. James B. Irwin, *To Rule the Night*, 60–61, 63, 6, 71, 80, 11, 88.
61. Ibid., 10–11, 193; "Baptist Astronaut Tells Church He Felt Close to God on Moon," *Baptist Press*, 25 August 1971, Baptist Press Archives, Southern Baptist Historical Library and Archives, Nashville, www.sbhla.org/bp_archive/index.asp.
62. James B. Irwin, *To Rule the Night*, 193; "42,000 Attend Texas 'Spiritual Spectacular,'" *Baptist Press*, 28 October 1971, Baptist Press Archives.
63. James B. Irwin, *To Rule the Night*, 195–98; "Astronaut James Irwin Sets Up Independent Organization," *Baptist Press*, 30 June 1972, Baptist Press Archives.
64. Mary Irwin, *Moon Is Not Enough*, 129–30. Detailed plans for the spiritual retreat—High Flight Lodge—can be found in High Flight publicity material, undated, folder 12, container 24, Records of the International Christian Broadcasters, Collection 86, Billy Graham Center Archives.
65. Scott and Leonov, *Two Sides of the Moon*, 233, 236–37.
66. Ibid., 379.
67. Ibid., 379–80. See also Andrew Smith, *Moondust*, 288–89.
68. Aldrin, *Magnificent Desolation*, 61.
69. Ibid., 170–75.
70. White, *Overview Effect*, 212–13; Chaikin, *Man on the Moon*, 266.
71. Chaikin, *Man on the Moon*, 582–83.
72. Duke and Duke, *Moonwalker*, 258.
73. Ibid., 258–60.
74. James Gorman, "Righteous Stuff," *Omni*, May 1984, 100. See also Duke and

Duke, *Moonwalker*, 280; and Charles Duke, "The Adventure Goes On," *Guideposts*, July 1984, 6.

75. Anders's wife, Valerie, opposed the church's position on birth control. Zimmerman, *Genesis*, 293.

76. Cunningham, *All-American Boys*, 26–27, 45; Scott and Leonov, *Two Sides of the Moon*, 94, 237; Chaikin, *Man on the Moon*, 47–48.

77. Mitchell, *Way of the Explorer*, 29.

78. Ibid., 42.

79. Ibid., 41–42.

80. Chaikin, *Voices From the Moon*, 17.

81. James B. Irwin, *To Rule the Night*, 98–100.

82. In 1963, 80% of Southern Baptist church members attested to such an experience. Among the other major denominations, the next highest figure was 50%. Stark and Glock, *American Piety*, 130–31.

83. James, *Varieties of Religious Experience*, 329, 348–51.

84. Otto, *Idea of the Holy*, 2.

85. Ibid., 7.

86. Andrew Smith, *Moondust*, 133–36, 276–78.

87. "To Walk the Moon," *New York Times*, 20 July 1969, 12.

88. "Apollo 16 Lunar Surface Journal: Geology Station 4 at the Stone Mountain Cycnos," edited by Eric M. Jones, NASA History Division, www.hq.nasa.gov/alsj/a16/a16.html.

89. William Anders, quoted in *This Island Earth*, edited by Oran W. Nicks (Washington, DC: NASA, 1970), 14. For examples of the image's use by other astronauts, see Cernan and Davis, *Last Man on the Moon*, 316; Duke and Duke, *Moonwalker*, 50; James B. Irwin, *To Rule the Night*, 1; and Scott and Leonov, *Two Sides of the Moon*, 301.

90. Collins, *Carrying the Fire*, 472, emphasis in original.

91. Chaikin, *Man on the Moon*, 555–56.

92. Charles A. Lindbergh, *Spirit of St. Louis*, 463–64.

93. For an insightful comparative analysis of Lindbergh's memoir and those of the astronauts, see Robert Baehr, "The Moon: How Far Is It . . . to Charles Lindbergh?," in *American Studies in Transition*, edited by David E. Nye and Christen Cold Thomsen (Odense, Denmark: Odense University Press, 1985), 153–73.

94. Charles Lindbergh, foreword to Collins, *Carrying the Fire*, ix, xi.

95. Collins, *Carrying the Fire*, 408–9, 484–86.

96. Ibid., 471, 477.

97. Ibid., 483.

98. Zimmerman, *Genesis*, 293–95.

99. Chaikin, *Man on the Moon*, 47–48; Francis French and Colin Burgess, *In the Shadow of the Moon: A Challenging Journey to Tranquility, 1965–1969* (Lincoln: University of Nebraska Press, 2007), 332–36.

100. French and Burgess, *In the Shadow of the Moon*, 343–47, 354–57; Russell L. Schweickart, oral history interview by Rebecca Wright, 19 October 1999, NASA/JSC Oral History Project, 27–32, 38–40, www.jsc.nasa.gov/history/oral_histories/oral_histories.htm.

101. Schweickart, oral history interview, 19 October 1999, 27–32, 38–40.

102. "Space: The Greening of the Astronauts."

103. French and Burgess, *In the Shadow of the Moon*, 362; William Irwin Thompson, *Passages about Earth: An Exploration of the New Planetary Culture* (New York: Harper & Row, 1973).

104. William Irwin Thompson, *Passages about Earth*, 5–6, 187; Katz, Marsh, and Thompson, *Earth's Answer*.

105. Schweickart, "No Frames, No Boundaries."

106. French and Burgess, *In the Shadow of the Moon*, 362–63.

107. Stewart Brand, ed., *Space Colonies* (Harmondsworth, UK: Penguin, 1977), 110.

108. See, e.g., Katz, Marsh, and Thompson, *Earth's Answer*; Brand, *Space Colonies*; and "Rediscovering the North American Vision," ed. Robert Gilman, special issue, *In Context* 3 (Summer 1983): 16–18.

109. See Poole, *Earthrise*, 161–66; White, *Overview Effect*, 74; *One People, One Planet*, directed by David Hoffman (1989; Santa Cruz, CA: Varied Directions, 2004), DVD; and Chris Pales, "Universal Awe Inspires Meacham's 'Island in Space,'" *Los Angeles Times*, 27 October 1993, F2.

110. Russell L. Schweickart, interview by Frank White, 29 October 1985, in White, *Overview Effect*, 201; Poole, *Earthrise*, 114–15; ASE history, Association of Space Explorers, www.space-explorers.org/association/history.html.

111. Poole, *Earthrise*, 164–65.

112. Russell L. Schweickart, oral history interview by Rebecca Wright, 8 March 2000, NASA/JSC Oral History Project, 37–39, www.jsc.nasa.gov/history/oral_histories/oral_histories.htm.

113. Brand, *Space Colonies*, 34, 74–81.

114. Schweickart, oral history interview, 19 October 1999.

115. Mitchell, *Way of the Explorer*, 3, emphasis in original.

116. Ibid., 62.

117. Ibid., 71–72.

118. Edgar D. Mitchell, "Appendix: Experiments with Uri Geller," in *Psychic Exploration: A Challenge for Science*, edited by Edgar D. Mitchell and John White (New York: G. P. Putnam's Sons, 1974), 683–86; "The Magician and the Think Tank," *Time*, 12 March 1973, www.time.com. See also James Randi, *The Truth About Uri Geller* (Amherst, NY: Prometheus Books, 1982).

119. Mitchell, *Way of the Explorer*, 91; Jim Schnabel, *Remote Viewers: The Secret History of America's Psychic Spies* (New York: Dell, 1997), 138–43.

120. Thomas S. Kuhn, *The Structure of Scientific Revolutions* (Chicago: University of Chicago Press, 1962).

121. "Reaching Beyond the Rational," *Time*, 23 April 1973, www.time.com; Richard S. Westfall, "Newton and the Hermetic Tradition," in *Science, Medicine, and Society in the Renaissance: Essays to Honor Walter Pagel*, edited by Allen G. Debus (New York: Science History Publications, 1972)), 2:183–98; Frances A. Yates, *Giordano Bruno and the Hermetic Tradition* (Chicago: University of Chicago Press, 1964); Yates, *The Rosicrucian Enlightenment* (London: Routledge, 1972).

122. Edgar D. Mitchell, "Introduction: From Outer Space to Inner Space . . . ," in Mitchell and White, *Psychic Exploration*, 36. See also Marilyn Ferguson, *The Aquarian Conspiracy: Personal and Social Transformation in the 1980s* (London: Routledge & Kegan Paul, 1981).

123. Mitchell, *Way of the Explorer*, 72–74, 83–91, 5, 75.

124. Ibid., 131.

125. "History of IoNS," Institute of Noetic Sciences website, www.noetic.org/about/history.cfm.

126. Mitchell, *Way of the Explorer*.

127. Andrew Smith, *Moondust*, 47–52.

128. James B. Irwin, *To Rule the Night*, 11, 193; "42,000 Attend Texas 'Spiritual Spectacular.'"

129. James B. Irwin, *To Rule the Night*, 11; "James Irwin—Testimony."

130. James B. Irwin, *To Rule the Night*, 194.

131. "Astronaut James Irwin Sets Up Independent Organization"; James B. Irwin, *More Than Earthlings: An Astronaut's Thoughts for Christ-Centered Living* (Basingstoke, UK: Pickering & Inglis, 1984), 8.

132. "Astronaut James Irwin Sets Up Independent Organization."

133. James B. Irwin, *To Rule the Night*, 199; "James Irwin to Travel Widely on Round of Baptist Meetings," *Baptist Press*, 1 September 1972, Baptist Press Archives.

134. Joseph B. Underwood to Bill Rittenhouse, 27 June 1972, folder 4521-15, Collection AR 551-3, International Mission Board Executive Office Files, Southern Baptist Historical Library and Archives.

135. James B. Irwin, *To Rule the Night*, 199–203; Underwood to Baker J. Cauthen, "Visits of Colonel James B. Irwin to Japan and Korea," 25 October 1972, folder 4521-15, Collection AR 551-3, Southern Baptist Historical Library and Archives; "Irwin Calls on Korea's Chief during Baptist-Sponsored Trip," *Baptist Press*, 27 October 1972, Baptist Press Archives; Underwood to Cauthen, 1 November 1972, folder 4521-15, Collection AR 551-3, Southern Baptist Historical Library and Archives; Underwood to Cauthen, "Visit of James Irwin to Vietnam and the Philippines," 16 November 1972, ibid.

136. Underwood to Mike Dixon, 1 November 1972, folder 4521-15, Collection AR 551-3, International Mission Board Executive Office Files, Southern Baptist Historical Library and Archives.

137. Underwood to Cauthen, "Visit of James Irwin to Vietnam and the Philippines."

138. "Former Astronaut Irwin Is in Evangelistic Orbit," *St. Louis Post-Dispatch*, 16 January 1973, 10–11.

139. "Former Astronaut Plans Retreat for Vietnam POWs," *Baptist Press*, 12 February 1973, Baptist Press Archives.

140. James B. Irwin, *To Rule the Night*, 208. There is evidence that even before Irwin's heart attack, despite the success of the tour of Asia, New Zealand, and Australia, the Foreign Mission Board had begun to have doubts about its relationship with the High Flight Foundation. In February 1973 Underwood complained that Rittenhouse had failed to contact him in time to confirm plans for a proposed FMB-sponsored visit to Europe in May and that the foundation, indeed, seemed to be trying to make its own arrangements for the visit, while expecting the board to cover its expenses. The board saw "no other alternative" but to postpone the trip. Underwood to James B. Irwin, 28 February 1973, folder 4521-14, Collection AR 551-3, Southern Baptist Historical Library and Archives.

141. Mary Irwin, *Moon Is Not Enough*, 130–31; "POW and MIA Retreats Draw 'High Flight' Support," *Baptist Press*, 24 July 1973, Baptist Press Archives; "POW-MIA Families Told to be Good Neighbors," ibid., 25 August 1973, Baptist Press Archives.

142. One newspaper, reporting Irwin's address to a rally of Missouri Baptists,

observed that he "did not mention a single contemporary moral issue, personal or social." See "Former Astronaut Irwin Is in Evangelistic Orbit." On Irwin's family life after his return from the moon, including details of continuing marital strain and problems with their children, see Mary Irwin, *Moon Is Not Enough*, 132–75. On the Southern Baptist Convention, see David T. Morgan, *The New Crusades, The New Holy Land: Conflict in the Southern Baptist Convention, 1968–1991* (Tuscaloosa: University of Alabama Press, 1996).

143. John C. Whitcomb and Donald B. DeYoung, *The Moon: Its Creation, Form and Significance* (Winona Lake, IN: BMH Books, 1978).

144. Larry Eskridge, "A Sign for an Unbelieving Age: Evangelicals and the Search for Noah's Ark," in *Evangelicals and Science in Historical Perspective*, edited by David N. Livingstone, D. G. Hart, and Mark A. Noll (Oxford: Oxford University Press, 1999), 255; James B. Irwin, *More Than an Ark on Ararat: Spiritual Lessons Learned While Searching for Noah's Ark*, with Monte Unger (Nashville: Broadman, 1985).

145. Gorman, "Righteous Stuff,", 100.

146. John Noble Wilford, "Obituary: James B. Irwin, 61, Ex-Astronaut; Founded Religious Organization," *New York Times*, 10 August 1991, 26.

147. James B. Irwin, *To Rule the Night*, 8.

148. Andrew Smith, *Moondust*, 229.

149. On the anticlimactic experience of lift-off, see Aldrin, *Return to Earth*, 219–20; Borman, *Countdown*, 201; Cunningham, *All-American Boys*, 123; and Schweickart, "No Frames, No Boundaries," 5. But not all of the astronauts found the launch process so smooth: see Collins, *Carrying the Fire*, 368–70; and Duke and Duke, *Moonwalker*, 18–20.

150. Chaikin, *Man on the Moon*, 227.

151. Adams, *Education of Henry Adams*, 317–26.

152. Anne Morrow Lindbergh, *Earthshine* (New York: Harcourt, Brace & World, 1969), 15–18.

153. Mailer, *Fire on the Moon*, 81–82.

154. *CBS Evening News*, 16 July 1969, Vanderbilt Television News Archive.

155. Thomas O. Paine, memorandum for record, 17 July 1969, folder: "Chronological File: July 1969," box 23, Paine Papers; Bernard Weinraub, "Some Applaud as Rocket Lifts, but Rest Just Stare," *New York Times*, 17 July 1969, 21; *CBS Evening News*, 16 July 1969, Vanderbilt Television News Archive.

156. Ayn Rand, "Apollo 11," *Objectivist*, September 1969, 5.

157. Tom Buckley, "Caribbean Cruise Attempts To Seek Meaning of Apollo," *New York Times*, 12 December 1972, 49, 53.

158. William Irwin Thompson, *Passages about Earth*, 1–9.

159. Isaac Asimov, *The Tragedy of the Moon* (New York: Dell, 1973), 212–13, emphasis in original.

160. *CBS Evening News*, 16 July 1969, Vanderbilt Television News Archive.

161. Barbara Marx Hubbard, *The Hunger of Eve: One Woman's Odyssey toward the Future* (Eastsound, WA: Island Pacific NW, 1989), 93.

162. David E. Nye, *Narratives and Spaces: Technology and the Construction of American Culture* (Exeter, UK: University of Exeter Press, 1997), 153.

163. John Updike, *Rabbit Redux* (Harmondsworth, UK: Penguin, 1973), 12.

164. Denis Cosgrove, *Apollo's Eye: A Cartographic Genealogy of the Earth in the Western Imagination* (Baltimore: Johns Hopkins University Press, 2001).

165. Stewart Brand, "The First Whole Earth Photograph," in Katz, Marsh, and Thompson, *Earth's Answer*, 185–88.

166. Poole, *Earthrise*, 35.

167. MacLeish, "Reflection."

168. Billy Watkins, *Apollo Moon Missions: The Unsung Heroes* (Westport, CT: Praeger, 2006), 30–31; Poole, *Earthrise*, 23–24.

169. Richard W. Underwood, oral history interview by Summer Chick Bergen, 17 October 2000, NASA/JSC Oral History Project, 51, www.jsc.nasa.gov/history/oral_his tories/oral_histories.htm.

170. Poole, *Earthrise*, 84–87.

171. Two weeks before the launch of *Apollo*, the United States Information Service described plans for telecasts during the flight, noting that audiences would be shown "the cloud-swirled ball of earth." This "should be rather spectacular," one NASA mission director declared. "Apollo Eight Astronauts Plan Six Telecasts during Moon Flight," United States Information Service, 6 December 1968, FCO 55/351, National Archives, Kew.

172. Poole, *Earthrise*, 198.

173. *NBC Nightly News*, 22 April 1970.

174. Collins, *Carrying the Fire*, 480–81.

175. *Life*, 28 March 1969, 57.

176. Byrnes, *Politics and Space*, 50–52; Kauffman, *Selling Outer Space*, 32–45.

177. Chaikin, *Man on the Moon*, 490; Allen, *Live from the Moon*, 177–80.

178. "Excerpts from the Apollo 8 Crew's News Conference," *New York Times*, 10 January 1969, 30. A year or so later, prior to his departure aboard *Apollo 13*, Lovell prepared a commentary on the moon to be delivered in lunar orbit. The commentary expressed regret that the *Apollo 8* astronauts had "fostered" the impression that the moon was monotonous: "We left the emphasis off the variety of structures we can see below us which match an Earth that had never seen water." But *Apollo 13* never made it into lunar orbit, and the commentary was not delivered. Apollo 13 Crew Log, box 6, Apollo Crew Logs, 1967–72, Flight Operations Directorate, Records of Johnson Space Center.

179. Daniel Dayan and Elihu Katz, *Media Events: The Live Broadcasting of History* (Cambridge, MA: Harvard University Press, 1994).

180. Zimmerman, *Genesis*, 246.

181. Kevin Michael Kertscher, "The Making of 'Race to the Moon': Apollo 8 Documentary Producer Tells All," *Ad Astra*, 20 October 2005, www.space.com/adastra.

182. "The View from Apollo 8," *Christian Century*, 8 January 1969, 37.

183. Richard M. Nixon, "Inaugural Address," 20 January 1969, *Public Papers of the Presidents: Richard M. Nixon, 1969*, 1–4.

184. "Notes and Comment," The Talk of the Town, *New Yorker*, 4 January 1969, 23.

185. Herbert E. Krugman, "Public Attitudes toward the Apollo Space Program, 1965–1975," *Journal of Communication*, Autumn 1977, 91.

186. Louis Harris, "49% Oppose Moon Project," *Philadelphia Inquirer*, 11 February 1969, 7.

187. Krugman, "Public Attitudes toward the Apollo Space Program," 90–91; Roger D. Launius, "Public Opinion Polls and Perceptions of US Human Spaceflight," *Space Policy* 19 (2003): 167–68.

188. Nye, *Narratives and Spaces*, 157.
189. Allen, *Live from the Moon*, 157.
190. Andrew Smith, *Moondust*, 9.
191. Mailer, *Fire on the Moon*, 92, 101 (quotation), 105.
192. J. Anthony Lukacs, "Some Random Events in the Nation on the Weekend It Put 2 Men on the Moon," *New York Times*, 21 July 1969, 8.
193. "Reactions to Man's Landing on the Moon Show Broad Variations in Opinions," ibid., 21 July 1969, 6–7.
194. "Giant Leap for Mankind?," *Ebony*, September 1969, 58.
195. "Blacks and Apollo," *New York Times*, 27 July 1969, 6.
196. Fifty-three percent of respondents answered that they opposed the United States' setting aside money for an effort to land a man on Mars. Gallup poll, 7 August 1969, in Gallup, *Gallup Poll*, 3:2209.
197. "Apollo XII Moon Landing and Follow-on Schedule," 18 November 1969, folder: "[Ex] OS 3 10/1/69–11/30/69," box 7, Subject Files: Outer Space, White House Central Files, Nixon Presidential Materials, National Archives, College Park.
198. "Apollo 12 Lunar Surface Journal: The First EVA: TV Troubles," edited by Eric M. Jones, NASA History Division, http://history.nasa.gov/alsj/frame.html.
199. Kraft, *Flight*, 330–31.
200. Lovell and Kluger, *Apollo 13*, 95.
201. Harlow to Haldeman, 14 April 1970; William R. MacKaye, "350 Pray for Apollo in District," *Washington Post*, 16 April 1970, 11. See also *Apollo 13: Houston, We've Got a Problem*, directed by Don Wiseman (1972; Houston, TX: A-V Corporation for NASA, 1972), internet video, www.archive.org/details/HoustonWeveGotAProblem.
202. Richard Nixon, "National Day of Prayer and Thanksgiving: A Proclamation," 17 April 1970, folder: "Reading File: April 1970," 1 of 2, box 36, Paine Papers.
203. "Apollo's Return: Triumph Over Failure," *Time*, 27 April 1970, www.time.com.
204. Krugman, "Public Attitudes toward the Apollo Space Program," 90–91.
205. Michael A. G. Michoud, *Reaching for the High Frontier: The American Pro-Space Movement, 1972–84* (New York: Praeger, 1986), 10–12, 41–45; Bainbridge, *Spaceflight Revolution*, 158–97.
206. Goldstein, *Flying Machine and Modern Literature*, 191–92; Karen Yelena Olsen, introduction to Olsen, *On the Wing*, 7.
207. Archibald MacLeish, "Voyage to the Moon," *New York Times*, 21 July 1969, 1.
208. Updike, *Rabbit Redux*, 348.
209. Mailer, *Fire on the Moon*, 379–80.
210. Goldstein, *Flying Machine and Modern Literature*, 207.
211. James Dickey, *Sorties* (Garden City, NY: Doubleday, 1971), 5, 48, emphasis in original.
212. Ibid., 55–56; Henry Hart, *James Dickey: The World As a Lie* (New York: Picador, 2000), 402–4.
213. James Dickey, "A Poet Witnesses a Brave Mission," *Life*, 1 November 1968, 26.
214. James Dickey, untitled poem, ibid., 10 January 1969, 23–26.
215. James Dickey, "The Moon Ground," ibid., 4 July 1969, 16C.
216. Joseph Campbell, *The Inner Reaches of Outer Space: Metaphor as Myth and as Religion* (New York: Harper & Row, 1986), 28.
217. Sagan, *Cosmic Connection*, 189–90.

CHAPTER FIVE: Evil Triumphs When Good Men Do Nothing

1. Richard M. Nixon to Spiro Agnew, Melvin R. Laird, Thomas O. Paine, and Lee A. Dubridge, 13 February 1969, folder: "[EX] FG 164: NASA: Beginning—8/3/69," box 1, Subject Files: FG 164: NASA, White House Central Files, Nixon Presidential Materials, National Archives, College Park.

2. Bryce Nelson, "Go for God," *Washingtonian Magazine*, November 1969, enclosed in Paine to Agnew, 31 October 1969, "Chronological File: Oct 1969," box 23, Paine Papers.

3. Paine to Agnew, 31 October 1969, "Chronological File: Oct 1969," box 23, Paine Papers.

4. One NASA document offers a total mail count of 4,268,396 for the years 1969–73, with 3,489,368 items listed under the heading "Scripture." "Mail Count 1969–1973," folder: "Complaints regarding the Reading of Selections from Genesis on Apollo 8," box 12, Apollo General History Series, JSC History Collection, University of Houston–Clear Lake.

5. "There's just no room to keep them all," said one NASA official in 1975. "O'Hair Rumor Blazes," *Austin American-Statesman*, 13 July 1975, 14; see also James A. Long, "Project Astronaut," 6 October 1969, Record No. 6744, NASA History Office. The largest surviving petition—with approximately 35,000 names—is stored on microfilm in the personal papers of Thomas O. Paine. See "The Lynn Petition," 6 February 1969, microfilm reel no. 1, box 48, Paine Papers.

6. *Christian Century*, indeed, editorialized: "We cannot help feeling that there are more serious issues confronting the Republic just now than this one." "Separationists Split Over Space Prayer," *Christian Century*, 19 November 1969, 1474. If there was an exception to this pattern, it lay in the efforts of the always exceptional Reverend Carl McIntire, president of the International Council of Christian Churches, but McIntire campaigned more vigorously to have the Bible represented on the federal stamp commemorating the flight of *Apollo 8* than he did to defend the rights of astronauts to actually read from it while in space. See *Christian Beacon*, 13 February, 6 March 1969.

7. A survey was conducted of tapes of broadcasts held at the Vanderbilt Television News Archive, Nashville. For examples of reports in the major newspapers, see "Letters Back Use of Bible by Astronauts," *Los Angeles Times*, 26 September 1969, 11; and "Prayer in Space Backed," *New York Times*, 19 December 1969, 41.

8. See esp. Peter Berger, *Sacred Canopy*; and Harvey Cox, *The Secular City: Secularization and Urbanization in Theological Perspective* (London: SCM, 1966). For useful secondary accounts, see Ellwood, *Sixties Spiritual Awakening*; and Hudnut-Beumler, *Looking for God*.

9. Alan Brinkley, "The Problem of American Conservatism," *American Historical Review* 99 (1994): 409–30; Leo P. Ribuffo, "Why is There So Much Conservatism in the United States and Why Do So Few Historians Know Anything about It?," ibid., 438–49.

10. Butler, "Jack-in-the-Box Faith." See also Paul Boyer, "In Search of the Fourth 'R': The Treatment of Religion in American History Textbooks and Survey Courses," *History Teacher* 29 (1996): 195–216; and Simon P. Newman, "One Nation Under God: Making Historical Sense of Evangelical Protestantism in Contemporary American Politics," *Journal of American Studies* 41 (2007): 581–97.

11. Currently, the most useful accounts of such activities are contained in Martin, *With God on Our Side*; and Williams, *God's Own Party*; with respect to school prayer, in Joan Delfattore, *The Fourth R: Conflicts Over Religion in America's Public Schools* (New Haven, CT: Yale University Press, 2004), 127–43; and with respect to creationism, in John A. Moore, "Creationism in California," *Daedalus* 103 (1974): 173–89; Dorothy Nelkin, *The Creation Controversy: Science or Scripture in the Schools* (New York: Norton, 1982); and Numbers, *Creationists*. For a vivid account of the development of grass-roots evangelical politics in Southern California between the Second World War and Ronald Reagan's victory in the 1980 presidential election, see Dochuk, *From Bible Belt to Sun Belt*.

12. See, e.g., Dan T. Carter, *The Politics of Rage: George Wallace, the Origins of the New Conservatism, and the Transformation of American Politics* (Baton Rouge: Louisiana State University Press, 2000); Thomas Byrne Edsall, *Chain Reaction: The Impact of Race, Rights, and Taxes on American Politics*, with Mary D. Edsall (New York: Norton, 1993); and Jonathan M. Schoenwald, *A Time for Choosing: The Rise of Modern American Conservatism* (Oxford: Oxford University Press, 2002). A significant exception to this trend can be found in the work of Lisa McGirr on Orange County, California, where right-wing politics were forcefully shaped by the strength of local conservative and, increasingly, evangelical Protestant congregations. Over time, indeed, the moral issues emphasized by these congregations—pornography, sex education, and abortion—came to claim a measure of precedence over economic libertarianism in the ideology of Orange County conservatives. McGirr, *Suburban Warriors*.

13. Jerome L. Himmelstein, "The New Right," in *The New Christian Right: Mobilization and Legitimation*, edited by Robert C. Liebman and Robert Wuthnow (New York: Aldine, 1983), 25–26; Michael Lienesch, *Redeeming America: Piety and Politics in the New Christian Right* (Chapel Hill: University of North Carolina Press, 1993), 8.

14. Martin, *With God on Our Side*, 117–43; Williams, *God's Own Party*, 105–32; Paul Boyer, "The Evangelical Resurgence in 1970s American Protestantism," in *Rightward Bound: Making America Conservative in the 1970s*, edited by Bruce J. Schulman and Julian E. Zelizer (Cambridge, MA: Harvard University Press, 2008), 29–51; Steve Bruce, *The Rise and Fall of the New Christian Right: Conservative Protestant Politics in America, 1978–1988* (Oxford: Clarendon, 1990), 108–9; Donald T. Critchlow, *Phyllis Schlafly and Grassroots Conservatism: A Woman's Crusade* (Princeton, NJ: Princeton University Press, 2005), 215–38.

15. Joseph Crespino, "Civil Rights and the Religious Right," in Schulman and Zelizer, *Rightward Bound*, 90–105; Martin, *With God on Our Side*, 169–73; Robert Freedman, "The Religious Right and the Carter Administration," *Historical Journal* 48 (2005): 238–41; Williams, *God's Own Party*, 163–64.

16. Jeffrey K. Hadden, Anson Shupe, James Hawdon, and Kenneth Martin, "Why Jerry Falwell Killed the Moral Majority," in *The God Pumpers*, edited by Marshall Fishwick and Ray B. Browne (Bowling Green, OH: Bowling Green State University Popular Press, 1987), 101–15; Robert C. Liebman, "Mobilizing the Moral Majority," in Liebman and Wuthnow, *New Christian Right*, 49–73.

17. In the summer of 1968, for example, 98% of respondents to a Gallup survey declared a belief in God. Gallup, *Gallup Poll*, 3:2174. In January 1969, 14% of those polled favored increasing the funding for "space research," 41% wished to maintain it at the same level, and 40% asserted that it should be scaled back. Ibid., 2183–84. For

public attitudes toward the space program more generally, see Krugman, "Public Attitudes toward the Apollo Space Program"; Launius, "Public Opinion Polls"; and Nye, *Narratives and Spaces*, 147–60.

18. For evidence of the influence of such factors, see Gallup, *Gallup Poll*, 3:2183–84; Nye, *Narratives and Spaces*, 149–51; "Poll Shows Support for Nixon's 'First in Space Goal,'" *Space Business Daily*, 16 December 1968, in folder: "Reading File: Dec. 1968," box 32, Paine Papers; Louis Harris, "49% Oppose Moon Project"; and Julian Scheer to Paine, 10 September 1969, folder: "[EX] Outer Space [1969–1970]," 1 of 2, box 1, Subject Files: Outer Space, White House Central Files, Nixon Presidential Materials, National Archives, College Park.

19. Homer Bigart, "New Breed Astronauts," 1, 18.

20. Frank Borman to Paul F. Feigert, 25 April 1969, Record No. 37590, JSC History Collection, University of Houston–Clear Lake.

21. R. G. Rose, "Chronology of Borman Prayer on Apollo 8," 29 January 1969, Record No. 36532, JSC History Collection, University of Houston–Clear Lake.

22. George M. Low, "Special Notes for August 9, 1968, and Subsequent," 19 August 1968, in Logsdon, *Exploring the Unknown*, 7:690–702; Low, "Special Notes for November 10 and 11, 1968," 14 November 1968, in ibid., 706–10; Chaikin, *Man on the Moon*, 56–59; Kraft, *Flight*, 283–89. There were also concerns that if NASA held fast to the scheduled sequence of missions, the resulting delays might permit the Soviet Union to steal its thunder, at least by looping a manned craft around the moon and back to Earth. The lunar-orbital mission also offered an opportunity to test navigation and communications systems, as well as procedures for reentry at hypersonic speeds (the craft essentially would be falling back to Earth from a very great height).

23. Kranz, *Failure Is Not an Option*, 223–27; S. Neil Hosenball to William D. Ruckelshaus, 1 October 1969, "Civil 3502* (Madalyn O'Hair)," Records of US Attorney, Western District of Texas, San Antonio Division; "Apollo 8 Flight Journal," ed. Woods and O'Brien, "Day 1: Launch and Ascent to Earth Orbit," NASA History Division, http://history.nasa.gov/ap08fj/index.htm.

24. Borman, *Countdown*, 194; Frank Borman, oral history interview by Catherine Harwood, 13 April 1999, NASA/JSC Oral History Project, www.jsc.nasa.gov/history/oral_histories/oral_histories.htm. In an article for an evangelical magazine a few months after the mission, Borman did not mention the conversation with Scheer but instead attributed the search for something to say during the Christmas Eve broadcast to his own personal wish to offer the massive audience "a broader prayer" than the one that was to be recorded for his church. Borman, "Message to Earth," 3–6.

25. Borman, "Message to Earth," 4; Frank Borman, Jim Lovell and Bill Anders, "Our Journey to the Moon," *Life*, 17 January 1969, 31; Zimmerman, *Genesis*, 232.

26. Zimmerman, *Genesis*, 232–36; Watkins, *Apollo Moon Missions*, 70–71.

27. Simon Bourgin to Borman, undated, folder: "Apollo 8 Flight Plan," box 1, Apollo 8 Flight Data File, Mission Operations, RG 255, Records of the National Aeronautics and Space Administration, National Archives Southwest Region, Fort Worth.

28. Borman, "Message to Earth," 4.

29. Frank Borman, affidavit, 6 October 1969, "Civil 3502* (Madalyn O'Hair)," Records of US Attorney, Western District of Texas, San Antonio Division.

30. Bourgin to Borman, undated; Borman, "Message to Earth," 5; Borman, oral history interview, 13 April 1999.

31. Apollo 8 Mission Report, February 1969, NASA History Division, http://history.nasa.gov/ap08fj/pdf/a08-missionreport.pdf.

32. Apollo 8 Onboard Voice Transcription, 173, NASA/JSC History Portal, www.jsc.nasa.gov/history/mission_trans/apollo8.htm.

33. Ibid., 183.

34. Ibid., 195–97. The conclusion of the Christmas Eve broadcast, including the Genesis reading, can be viewed at www.jsc.nasa.gov/history/mission_trans/apollo8.htm.

35. Kranz, *Failure Is Not an Option*, 246. See also Kraft, *Flight*, 300.

36. "Back to 'Oasis' Earth," *New York Times*, 26 December 1968, 36.

37. "View from Apollo 8."

38. "Head Man at NASA: Thomas Otten Paine," *New York Times*, 28 December 1968, 14. Norman Mailer expressed a similar thought after the flight of *Apollo 11*, albeit from the other side of the cultural battlefront. He described "his people" as "an army of outrageously spoiled children who cooked with piss and vomit while the Wasps were quietly moving from command of the world to command of the moon." Mailer, *Fire on the Moon*, 357.

39. "Joint Meeting of the Two Houses of Congress to Receive the Apollo 8 Astronauts"; Borman, *Countdown*, 224.

40. In 1966 forty-nine senators voted for the Dirksen Amendment, which declared the constitutionality of schools' "providing for or permitting the voluntary participation by students or others in prayer." Robert S. Alley, *School Prayer: The Court, the Congress, and the First Amendment* (Buffalo, NY: Prometheus, 1994), 166. In June 1963, 70% of respondents to a Gallup poll disapproved of the Supreme Court rulings. Gallup, *Gallup Poll*, 3:1837.

41. *CBS Evening News*, 26 May 1969, Vanderbilt Television News Archive.

42. *Engel v. Vitale*, 370 U.S. 421 (1962).

43. "Church-State Separation," *Christianity Today*, 20 July 1962, 29–30.

44. Leo Pfeffer, "Information Bulletin: The Becker Amendment," 15 February 1964, folder: "Becker Amendment, 1964," box 82, Records of the American Jewish Congress, I-77, American Jewish Historical Society, Center for Jewish History, New York.

45. *Abingdon School Dist. v. Schempp*, 374 U.S. 203 (1963); Bryan F. Le Beau, *The Atheist: Madalyn Murray O'Hair* (New York: New York University Press, 2003), 98–110. It helped that American atheists were rather modest in number and meek in spirit. O'Hair had few competitors in her quest for public preeminence. Jacoby, *Freethinkers*, 313–16.

46. "Woman Attacks Space Prayer from Apollo 8," *Odessa (TX) American*, 27 December 1968, 7, www.NewspaperArchive.com.

47. "Fights Spaceflight Religion," *New Castle (PA) News*, 28 December 1968, 2, www.NewspaperArchive.com.

48. See, e.g., "She Wants No Prayer from Space," *Washington Post*, 28 December 1968, 9.

49. See, e.g., Linda Strong to Paine, 27 December 1968, folder: "Subject File: Apollo Missions: Apollo 8," box 43, Paine Papers.

50. See Mrs. Robert Welch et al. to Nixon, 4 January 1969, folder: "[GEN] OS 3 Space Flight Begin 6/10/69–," Subject Files: Outer Space, box 10, White House Central Files, Nixon Presidential Materials, National Archives, College Park.

51. Frank J. Magliato to Scheer, 6 January 1969, Record No. 6744, NASA History Office.

52. See, e.g., "Space Prayer Backed," *Frederick (MD) Post*, 31 December 1968, 5, www.NewspaperArchive.com.

53. See, e.g., Mail Readers' Point of View, *Charleston (WV) Daily Mail*, 16 and 18 January 1969, www.NewspaperArchive.com.

54. "Lynn Petition." Estimate provided by Keith Halderman.

55. "Worship in Space Lauded," *Houston Post*, 8 March 1969, 14. Fry also presented her petition to Thomas Paine, at NASA headquarters in Washington, on 10 March. Paul Droste to Paine, 10 March 1969, folder: "Subject File: Apollo Missions: Apollo 8," box 43, Paine Papers.

56. "Right to Pray," *Brainerd (MN) Daily Dispatch*, 21 April 1969, 3, www.NewspaperArchive.com; Dwight L. Chapin to Steve Gabrielse, 11 August 1969, folder: "[GEN] OS 3 8/8/69–8/18/69," Subject Files: Outer Space, box 10, White House Central Files, Nixon Presidential Materials, National Archives, College Park.

57. Bill Mansdoerfer to Jerome R. Waldie, 17 June 1969, folder: "Project Astronaut," box 93, Jerome R. Waldie Papers, Bancroft Library, Berkeley.

58. Borman, "Message to Earth," 6. O'Hair made no mention of the mooted mail campaign during her weekly broadcasts on KTBC Radio in Austin, Texas. Indeed, she discussed the religion-in-space issue on only one of her broadcast in 1969—on 8 December. For broadcast transcripts, see Madalyn O'Hair, *What on Earth is an Atheist!* (Austin: American Atheist, 1969); and O'Hair, *An Atheist Speaks* (Austin: American Atheist, 1986), 145–50.

59. "God and Astronauts Debated," *Detroit News*, 13 January 1969, 1C–2C; "Why an Atheist Battles Churches," *Detroit Free Press*, 19 January 1969, 15; Madalyn O'Hair to John Mitchell, 31 January 1969, folder: "[GEN] RM Religious Matters RM Begin 2-28-69," Subject Files: Religious Matters, box 1, White House Central Files, Nixon Presidential Materials, National Archives, College Park. The same day, O'Hair also sent a letter to President Nixon urging him to investigate the wealth of the churches so that Americans could better consider the question of their financing by the state either through direct grants or tax exemptions. O'Hair to Nixon, 31 January 1969, ibid.

60. William H. Rehnquist to Chapin, 18 February 1969, folder: "[GEN] RM Religious Matters RM Begin 2-28-69," Subject Files: Religious Matters, box 1, White House Central Files, Nixon Presidential Materials, National Archives, College Park; "Apollo Crewmen Didn't Quote Bible and Madalyn Murray is 'So Pleased,'" *Holland (MI) Evening Sentinel*, 10 March 1969, 18, www.NewspaperArchive.com.

61. *Christian Beacon*, 12 June 1969, 5.

62. Le Beau, *Atheist*, 147–48.

63. O'Hair et al., "Complaint and Prayer (excuse the expression) for Injunctive Relief," 5 August 1969, "Civil 3502* (Madalyn O'Hair)," Records of US Attorney, Western District of Texas, San Antonio Division.

64. According to Loretta Lee Fry in March 1969, O'Hair had "threatened to petition NASA with 28,000 signatures." The organizers of Project Astronaut asserted that O'Hair "has more than 27,000 signatures." In May 1973 the Vienna, Virginia, United Methodist Church bulletin noted that "now she has obtained 27,000 signed letters" protesting the *Apollo 8* Genesis reading. "Worship in Space Lauded"; "Project Astronaut: What Each One Can Do," undated, folder: "Reading File: Sept. 1969," 2 of 2, box

34, Paine Papers; bulletin enclosed in Gerald J. Mossinghoff to Frank X. Krebs, 27 April 1973, Record No. 6742, NASA History Office.

65. "O'Hair Rumor Blazes."

66. Le Beau, *Atheist*, 110–14, 125–29.

67. Madalyn Murray, interview by Richard Tregaskis, *Playboy*, October 1965, www.positiveatheism.org/hist/madplay.htm, emphasis in original.

68. *Life*, 12 April 1963, 63.

69. Long, "Project Astronaut."

70. "Right to Pray"; "Prayer in Space Backed"; "Mrs. Fry Enroute to NASA with 500,000 'Prayer' Signatures," *Detroit News*, 6 March 1969, 3D.

71. "Jaycees Get Lots of Names," *Big Spring (TX) Herald*, 7 October 1969, 6.

72. See, e.g., Strong to Paine, 27 December 1968; and "Sunday School Members Commend Astronaut Trio," *Las Cruces (NM) Sun News*, 17 March 1969, 9.

73. For examples, see a petition entitled "Thank You Astronauts!" collected in La Habra, CA, and sent to President Nixon in March 1968, folder: "[GEN] OS 3-1 Astronauts Begin—3/31/69," Subject Files: Outer Space, box 12, White House Central Files, Nixon Presidential Materials, National Archives, College Park; a petition circulated in Antioch, CA, in August 1969, and sent to Congressman Jerome R. Waldie of California's Fourteenth District, enclosed in Kathleen J. Anderson to Waldie, undated, folder: "O'Hair Criticism re: God," box 93, Waldie Papers; a letter received by Congressman Donald E. Lukens of Ohio's Twenty-fourth District: Lester C. Owen to Lukens, 5 March 1969, folder "O," box 7, Donald E. Lukens Papers, Ohio Historical Society, Columbus; a petition collected in Greensburg, IN, and sent to US District Attorney Seagal Wheatley, enclosed in W. M. Cook to Wheatley, 19 September 1969, "Civil 3502* (Madalyn O'Hair)," Records of US Attorney, Western District of Texas, San Antonio Division; and seventeen letters on the subject of O'Hair and *Apollo 8* published in Sound-off, *Houston Post*, 3 January 1969, sec. 3, p. 3.

74. "Mail Count 1969–1973"; "O'Hair Rumor Blazes"; Fletcher to Wallace F. Bennett, 31 August 1973, Record No. 6742, NASA History Office.

75. Harold Camping to Paine, 24 September 1969, folder: "Reading File: Sept 1969," 2 of 2, box 34, Paine Papers.

76. This total, of course, cannot be viewed as entirely reliable. The figures for the Project Astronaut, Gabrielse, Fry, and Jaycees petitions all came from the petitioners themselves; in 1969, at least, NASA conducted no independent audits. Many Americans may have attached their name to more than one letter or petition. It is also conceivable that the 646,000 signatures submitted by the Gabrielses in mid-December 1969 may have been included in the NASA mail count for 1970. Yet there are reasons for considering the total to be an undercount of the level of correspondence activity, most obviously because it excludes letters and petitions sent to destinations other than NASA. Although no figures have been found for 1974, a goal of securing 1 million signed letters commending the Genesis reading was set that year by the American Legion in California. Virgil C. Dechant to Fletcher, 16 September 1974, folder: "Correspondence (1 Aug–19 Sept 1974)," box 24, Fletcher Papers. In addition, it is unclear whether the figures collected by NASA were for individual items of mail or for the number of signatures that each item contained. Finally, the figure of 1 million for 1975 was reported in mid-July, following a two-week period in which 200,000 items of mail

had been received. It is unlikely that the influx ended suddenly thereafter. "O'Hair Rumor Blazes."

77. Dierenfield, *Battle over School Prayer*, 151, 179–80.

78. Ibid., 154–55; John F. Kennedy, "The President's News Conference of June 27th, 1962," in *The American Presidency Project*, edited by John T. Woolley and Gerhard Peters, www.presidency.ucsb.edu/ws/?pid=8735. See also Steven K. Green, "Evangelicals and the Becker Amendment: A Lesson in Church-State Moderation," *Journal of Church and State* 33 (1991): 541–67.

79. Minutes of meeting of Board of Directors, ACLU Central Texas Chapter, 13 October 1969, folder 3, box 21, Fort Worth Civil Liberties Union Papers (AR108), Special Collections, University of Texas at Arlington Library; Melvin L. Wulf to Roy M. Mersky, 4 November 1969, folder: "O'Hair, Madalyn M., et al., v. Paine, Thomas O.," box 1596, American Civil Liberties Union Papers, Seeley G. Mudd Manuscript Library, Princeton University, Princeton, NJ.

80. Franklin C. Salisbury to Wheatley, 16 December 1969, "Civil 3502* (Madalyn O'Hair)," Records of US Attorney, Western District of Texas, San Antonio Division; "Outer Space Prayer Ban Held Absurd," *Los Angeles Times*, 19 January 1969, G7; Joseph B. Robison, "Religion and Government: Questions and Answers," undated, folder: "Church and State: Miscellaneous Publications, etc.," box 37, Records of the American Jewish Congress, I-77. On O'Hair's relationship with the American Jewish Congress and Americans United, see Frank J. Sorauf, *The Wall of Separation: The Constitutional Politics of Church and State* (Princeton, NJ: Princeton University Press, 1976), 49, 107.

81. Fry, the Gabrielses, and the organizers of Project Astronaut all requested an appointment with President Nixon, hoping to present their petitions to him in person. All were turned down. See Charles S. Gubser to William Timmons, 2 July 1969, folder: "[GEN] OS 3-1 4/1/69–7/24/69," Subject Files: Outer Space, box 12, White House Central Files, Nixon Presidential Materials, National Archives, College Park; Harlow Shapley to Scheer, "Project Astronaut," 3 July 1969, Record No. 6744, NASA History Office; Chapin to Gabrielse, 11 August 1969.

82. "Madalyn Needs Padalyn," *Detroit Free Press*, 31 December 1968, 8. The exception was the sympathetic coverage of Loretta Lee Fry's petition campaign provided by the *Detroit News*. See "God and Astronauts Debated"; "Mrs. Fry Enroute to NASA"; and "Prayer Petitions Presented," *Detroit News*, 8 March 1969, 1.

83. Lukens to Owen, 25 March 1969, folder "O," box 7, Lukens Papers; Stuart Symington, "Christmas Eve in Space," Cong. Rec. 1252–53 (1969); Waldie to George P. Miller, 5 September 1969, folder: "Project Astronaut," box 93, Waldie Papers; Long, "Project Astronaut."

84. For the Ohio Jaycees campaign, see "Jaycees Want Outer Space Bible Reading," *Lima (OH) News*, 2 December 1969, 21, and "Prayer Ban Protest," ibid., 10 April 1970, 10, www.NewspaperArchive.com. For the American Legion in California, see Dechant to Fletcher, 16 September 1974.

85. "Lynn Petition." Estimate provided by Keith Halderman.

86. "Bible Broadcaster See 69's Hot Topic," *Detroit Free Press*, 4 January 1969, 6; "God and Astronauts Debated; "Mrs. Fry Enroute to NASA"; "Worship in Space Lauded"; Droste to Paine, 10 March 1969, folder: "Subject Files: Apollo Missions: Apollo 8," box 43, Paine Papers.

87. Camping to Paine, 24 September 1969.

88. For example, see undated clipping from the *Vienna (GA) News* reprinting an article originally published in the *Wesleyan Christian Advocate*, the official newspaper of the North and South Georgia Conferences of the United Methodist Church, folder: "Project Astronaut," box 93, Waldie Papers.

89. "Project Astronaut," *Los Angeles Times*, 13 July 1969, SG-A6.

90. See undated clipping in folder: "Project Astronaut," box 93, Waldie Papers.

91. Camping to Paine, 24 September 1969.

92. For examples, see "Lynn Petition."

93. Nelkin, *Creation Controversy*, 57–70, 93–119; Numbers, *Creationists*, 268–79, 329–50; Harold Lindsell, *The Battle for the Bible* (Grand Rapids, MI: Zondervan, 1978).

94. Mrs Frank Jessup to Paine and Mitchell, c. 11 August 1969, "Civil 3502* (Madalyn O'Hair)," Records of US Attorney, Western District of Texas, San Antonio Division.

95. Bigart, "New Breed Astronauts," 1, 18. It was probably in much the same spirit that Buzz Aldrin, during *Apollo 11*'s final telecast, quoted from the eighth psalm: "When I consider the heavens, the work of Thy fingers, the moon and the stars which Thou hast ordained, what is man that Thou art mindful of him." Apollo 11 Technical Air-to-Ground Voice Transcription, 590, Record No. 6744, NASA History Office.

96. "Space Prayer Backed."

97. Judy Dávalos to Waldie, 13 August 1969, folder: "O'Hair Criticism re: God," box 93, Waldie Papers, emphasis in original.

98. "Petition to NASA," 22 March 1970, folder: "[GEN] OS Outer Space [1967–1970]," Subject Files: Outer Space, box 2, White House Central Files, Nixon Presidential Materials, National Archives, College Park, emphasis in original.

99. Howard W. Robison, "Faith in Space," Cong. Rec. E554 (1969).

100. Mail Readers' Point of View, *Charleston (WV) Daily Mail*, 20 January 1969, www.NewspaperArchive.com.

101. "Jaycees Report Good Response," *Odessa (TX) American*, 29 September 1969, 2, www.NewspaperArchive.com.

102. "Apollo Crewmen Didn't Quote Bible."

103. Mail Readers' Point of View, *Charleston (WV) Daily Mail*, 20 January 1969.

104. "Project Astronaut: What Each One Can Do."

105. In early 1969 the federal government received sizable quantities of protest correspondence on two similar issues: first, in response to rumors that the phrase *In the beginning, God* had been removed from the design of the federal stamp commemorating the *Apollo 8* mission; second, following a news story that army chaplains had been ordered not to make reference to God while teaching in the service's character-guidance programs. Petition from the Community Church of Pine Grove, CA, 20 February 1969, folder: "[GEN] OS 3 Space Flight Begin 6/10/69–," Subject Files: Outer Space, box 10, White House Central Files, Nixon Presidential Materials, National Archives, College Park. *Christian Beacon*, 6 February 1969, 1, 8; 13 February 1969, 5; 6 March 1969, 1, 4–5. "Chaplains' Role Under New Scrutiny," *Christianity Today*, 25 April 1969, 32. Anne C. Loveland, "Character Education in the U.S. Army, 1947–1977," *Journal of Military History* 64 (2000): 810–12.

106. Camping to Paine, 24 September 1969.

107. "Worship in Space Lauded."

108. Haldeman diary entry, 14 July 1969, box 1, Handwritten Journals and Diaries of Harry Robbins Haldeman, Nixon Presidential Materials, National Archives, College

Park; "Prayers Offered at White House," *New York Times*, 21 July 1969, 10; "President's Proclamation," ibid., 17 July 1969, 22; Borman to Nixon, 14 July 1969, folder: "[EX] OS 3 7/1/69–7/21/69," box 4, Subject Files: Outer Space, White House Central Files, Nixon Presidential Materials, National Archives, College Park; "Transcript of Nixon's Talk on Carrier," *New York Times*, 25 July 1969, 29.

109. Julian Scheer, "What President Nixon Didn't Know," *Space.com*, 16 July 1999, www.space.com.

110. Aldrin, *Return to Earth*, 233.

111. J. W. Ould, "Legal Office: Weekly Activity Report—August 9–15, 1969," folder: "August 68," box 2, Weekly Activity Reports to NASA Administrator, Office of the Director, Records of the Johnson Space Center.

112. J. B. Medaris to Paine, 21 March 1969, Record No. 6744, NASA History Office; "Prepared Statement of Spencer M. Beresford," *Chapel of the Astronauts: H.R. 4545: Hearing Before the Subcommittee on Manned Space Flight*, Committee on Science and Astronautics, US House of Representatives, 92nd Cong. 9 (1971). NASA proposed instead first a long-term lease of the land and then its sale, but the chapel was never built.

113. Jack Roberts, memorandum opinion: *O'Hair v. Paine*, 1 December 1969, enclosed in James H. Anderson Jr. and O'Hair, "Petition for Writ of Certiorari to the United States Court of Appeals for the Fifth Circuit," 15 December 1970, "Civil 3502* (Madalyn O'Hair)," Records of US Attorney, Western District of Texas, San Antonio Division; US Court of Appeals for the Fifth Circuit, per curiam: *O'Hair v. Paine*, 22 September 1970, ibid.

114. See Apollo Prayer League newsletter, c. February 1971.

115. James B. Irwin, *To Rule the Night*, 61.

116. See, e.g., Ould to George S. Trimble et al., "Identification of information sources needed to prepare defense in the case of Madalyn Murray O'Hair, et al. v. Thomas O. Paine, et al.," 2 September 1969, Record No. 39345, JSC History Collection, University of Houston–Clear Lake; Hosenball to Ruckelshaus, 1 October 1969, "Civil 3502* (Madalyn O'Hair)"; Edwin E. Aldrin Jr., affidavit, 26 September 1969, "Civil 3502* (Madalyn O'Hair)," Records of US Attorney, Western District of Texas, San Antonio Division.

117. See, e.g., George M. Low to George P. Shultz, 17 August 1971, folder: "[EX] OS 3 9/1/71–10/31/71," box 8, Subject Files: Outer Space, White House Central Files, Nixon Presidential Materials, National Archives, College Park; Space Task Group, *The Post-Apollo Space Program: Directions for the Future*, September 1969, folder: "[EX] FG 221-18 Space [1969–70]," box 3, FG 222: Task Forces, Subject Files, White House Central Files, Nixon Presidential Materials, National Archives, College Park.

118. Philip H. Abelson, "Apollo and Post-Apollo," *Science*, 10 October 1969, 171; Arnold S. Levine, *Managing NASA in the Apollo Era* (Washington, DC: NASA, 1982), 182; John M. Logsdon, "From Apollo to Shuttle: Policy Making in the Apollo Era," spring 1983, typescript, chap. 4, 53–55, NASA History Office.

119. Fletcher to Donnelly, 1 March 1973, folder: "Correspondence (15 Jan–24 April 1973)," box 16, Fletcher Papers; Fletcher to Donnelly, 17 May 1974, Record No. 6742, NASA History Office. For a useful article on Fletcher, see Launius, "Western Mormon in Washington, D.C."

120. Donnelly to Alfred P. Alibrando, "Ideas from the Field on How to Reach the Religious Market," 9 October 1973, Record No. 6742, NASA History Office.

121. "Project America: The Silent Majority Speaks," c. September 1969, folder: "Reading File: Sept 1969," 2 of 2, box 34, Paine Papers.

122. "Mail Fearing Religious-Shows Ban Deluges F.C.C.," *New York Times*, 18 February 1976, 74; Federal Communications Commission, "FCC 75-946: Memorandum Opinion and Order," 1 August 1975, ftp://ftp.fcc.gov/pub/Bureaus/Mass_Media/Databases/documents_collection/75-946.pdf.

123. "Mail Fearing Religious-Shows Ban Deluges F.C.C."; Alan Emory, "God Still on Air," *Pasadena Star-News*, 5 November 1975, 12, www.NewspaperArchive.com.

124. See, e.g., "O'Hair Move Discussed by Ministers," *Monessen Valley (PA) Independent*, 16 August 1976, 2; and "Really, Mrs O'Hair isn't in Walla Walla," *Walla Walla (WA) Union-Bulletin*, 14 January 1977, 11, www.NewspaperArchive.com.

125. "Mail Fearing Religious-Shows Ban Deluges F.C.C."; "Really, Mrs O'Hair Isn't in Walla Walla."

126. "Mail Fearing Religious-Shows Ban Deluges F.C.C."

127. Jim Castelli, "The Curse of the Phantom Petition," *TV Guide*, 24–30 July 1976, 4–5.

128. Federal Communications Commission, "FCC 75-946: Memorandum Opinion and Order."

129. Quentin J. Schultze, "Keeping the Faith: American Evangelicals and the Media," in Schultze, *American Evangelicals and the Mass Media*, 31–34.

130. Ben Armstrong, "RM 2493 Breaks All-Time FCC Record," *Religious Broadcasting* 7 (Summer 1975): 2–4; LaVay Sheldon, "Special Report Part II," ibid. 26.

131. Charles Secrest, "Yes . . . She's at It Again!," *Christian Crusade Weekly*, 15 June 1975, 1–2.

132. Charles Secrest, "But I'm Not a Fighter . . . ," ibid., 20 July 1975, 1.

133. Ibid.; Letters From Our Readers, *Albert Lea (MN) Evening Tribune*, 12 August 1975, 4.

134. "The Petition Against God," *Channels of Communication* 4 (September–October 1984): 10–14; "O'Hair Myth Lives On," *Dallas Morning News*, 25 January 1986, 88.

135. "Irwin Defends Space Costs; Cites Christian Testimony," *Baptist Press*, 8 June 1972, Baptist Press Archives.

Epilogue

1. Burrows, *This New Ocean*, 556.

2. For details of the Teacher In Space Program and McAuliffe's scheduled activities, see Colin Burgess, *Teacher In Space: Christa McAuliffe and the Challenger Legacy* (Lincoln: University of Nebraska Press, 2000); and NASA, "Space Shuttle Mission STS-51L: Press Kit," January 1986, NASA History Division, http://history.nasa.gov/sts51l/presskit.pdf.

3. Burgess, *Teacher in Space*, 54–55.

4. Peggy Noonan, *What I Saw at the Revolution: A Political Life in the Reagan Era* (New York: Random House, 1990), 253–57; Mary E. Stuckey, *Slipping the Surly Bonds: Reagan's Challenger Address* (College Station: Texas A&M University Press, 2006), 61–80.

5. Noonan, *What I Saw at the Revolution*, 258–59.

6. Collins, *Carrying the Fire*, 245.

7. Greg Klerkx, *Lost in Space: The Fall of NASA and the Dream of a New Space Age* (London: Secker & Warburg, 2004), 24–26; Burrows, *This New Ocean*, 554–61.

8. Burgess, *Teacher in Space*, 56.
9. See Collins, *Carrying the Fire*, 52.
10. Klerkx, *Lost in Space*, 177–79.
11. See George W. Bush, "Address to the Nation on the Loss of Space Shuttle *Columbia*," 1 February 2003, in Woolley and Peters, *American Presidency Project*, www.presidency.ucsb.edu/ws/?pid=181.
12. Peggy Noonan, "The Days of Miracle and Wonder," *Wall Street Journal*, 1 February 2003, http://online.wsj.com/article/SB122427328139145395.html.

Bibliographic Essay

In the social rounds of academic life, at seminars, symposia, and conferences, an introduction to and a handshake with a stranger are usually followed by the question, "So, what are you working on?" Over the course of researching and writing this book, my answer to that question has tended to elicit one of two responses. The first, a slightly blank stare and the follow-up, "What do religion and the space program actually have to do with one another?" The second, "I can't believe nobody has worked on that before." These responses, though contrasting, reflect a shared surmise: that the subject has not been productive of a literature. In truth, a literature—both primary and secondary—exists, but it is diffuse rather than conspicuous.

The most substantial published analysis of what space travel has historically owed to religion is contained in David F. Noble's *The Religion of Technology: The Divinity of Man and the Spirit of Invention* (New York: Knopf, 1997). Noble argues that space technology exemplifies mankind's desire to recover the divinity that it lost in the Fall. In a PhD dissertation completed at the University of Texas at Austin in 2004, "Space Rapture: Extraterrestrial Millennialism and the Cultural Construction of Space Colonization," Ryan McMillen advanced a similar, if not identical, thesis: that the impulse toward space travel was inspired by the Christian apocalyptic, expressing a rapturous fantasy of escape from an annihilated earth. As I make clear in chapter 1, I disagree with Noble and McMillen, though my own argument has been sharpened by their provocations. Both authors draw substantially on a voluminous file of documents under the heading "Religion" collected by staff at the NASA History Office in Washington, DC. A survey of that file naturally leads to the conclusion that space exploration's relationship to religion was consistent and intimate, because all the material it contains attests to that relationship. But this constitutes a sampling error. The operations of faith needed to be examined in context. In the course of my research for this book, I scrutinized the office files of senior NASA personnel, including Wernher von Braun (Library of Congress), Thomas Paine (Library of Congress), Homer Newell (National Archives, College Park), and James Fletcher (University of Utah), as well as the records of the Johnson Space Center in Houston (National Archives Southwest Region, Fort Worth, and University of Houston–Clear Lake). This was a rather different research experience, an immersion in day-to-day institutional logics that were secular in essence and usually drably instrumental. In *The Spaceflight Revolution: A Sociological Study* (New York: John Wiley & Sons, 1976), William Sims Bainbridge also reports that spaceflight was sometimes discussed by its advocates in religious terms, particularly as a source of peak experience, but he does not seem to regard religion as a determinative influence in the making of the space age. His principal case study, Barbara Marx Hubbard's Committee for the Future, emerged only after the first two moon landings; moreover, within a few years the committee had abandoned its focus on space.

Religion may not have been a primary motivation for most of those developing the means to travel into space, or for the astronauts who would actually make the journey, but it did provide an important frame within which they and others could understand what they were doing. Robert Poole's *Earthrise: How Man First Saw the Earth* (New Haven, CT: Yale University Press, 2008) usefully records how commonly contemporary discussions of spaceflight made reference to its religious significance and how some of the astronauts responded spiritually to their encounter with the cosmos. However, as indicated in chapter 3 of this book, I am less persuaded than Poole that what charged the experiences of the astronauts with religious meaning was always primarily the view of the earth. For the most extensive—though still not very extensive—discussion of the space age from the perspective of a religious historian, see Robert S. Ellwood's accounts of American religion in the fifties and sixties: *The Fifties Spiritual Marketplace: American Religion in a Decade of Conflict* (New Brunswick, NJ: Rutgers University Press, 1997) and *The Sixties Spiritual Awakening: American Religion Moving from Modern to Postmodern* (New Brunswick, NJ: Rutgers University Press, 1994).

The most detailed historical survey of modern astrotheology is contained in Steven J. Dick, *Life on Other Worlds: The 20th-Century Extraterrestrial Life Debate* (Cambridge: Cambridge University Press, 1998). Dick, however, is principally interested in speculations about the theological significance of extraterrestrial life, and as I explain in chapter 2, theological assessments of man's entry into the heavens during the space age ranged more widely than that. Many of these assessments took the form of short articles dispersed across religious or theological journals or brief digressions buried in the body of more general theological monographs. The fullest contemporary consideration of what the space age might mean for modern theology can be found in Martin J. Heinecken, *God in the Space Age* (Philadelphia: Winston, 1959).

No historian of the space program suffers from a shortage of participant accounts. For the exploration of religious experience in spaceflight contained in chapter 3, I draw on a number of published memoirs, particularly Edgar Mitchell, *The Way of the Explorer: An Apollo Astronaut's Journey through the Material and Mystical Worlds*, with Dwight Williams (New York: G. P. Putnam's Sons, 1996), and James B. Irwin, *To Rule the Night*, with William A. Emerson Jr. (Old Tappan, NJ: Spire Books, 1975). Also valuable were the interviews conducted since 1996 by the Johnson Space Center Oral History Project, all of which are available online.

The principal resource for the account in chapter 4 of Madalyn Murray O'Hair's legal challenge to religious speech in space were the records of the US attorney in Texas, who was responsible for preparing the government's defense (National Archives Southwest Region, Fort Worth). Material relating to the correspondence campaigns supporting the Apollo 8 Genesis reading was gathered from a range of different collections, including the NASA History Office, the records of the Johnson Space Center held at the University of Houston–Clear Lake, the White House Central Files in the Nixon Presidential Materials (then located at the National Archives, College Park, and now at the Nixon Presidential Library in Yorba Linda, California), and the papers of James Fletcher (University of Utah) and Thomas Paine (Library of Congress). The Paine Papers contain a microfilm of the Lynn petition, the only one of the major petitions that seems to have survived. It was also possible to track the grass-roots development of the correspondence campaigns by accessing the collection of local and regional newspapers available at www.NewspaperArchive.com.

Finally, throughout the book, the reader will find frequent reference to two monumental and highly individual journalistic inquiries into the space age: Oriana Fallaci's *If the Sun Dies* (New York: Atheneum, 1967) and Norman Mailer's *A Fire on the Moon* (London: Weidenfeld & Nicolson, 1970). Fallaci and Mailer were both secular writers, but they shared an intuition that ideas about God and his relationship with mankind were at work—and at stake—as mankind moved to set its sights on the moon.

Index

Abernathy, Ralph, 123
Abingdon School Dist. v. Schempp, 152
abortion, 140, 142
Adams, Henry, 82–83, 122
African Americans, 130
Agnew, Spiro, 137–38
Aldrin, Buzz, 58, 105, 112, 122, 129–30, 169; communion celebrated on moon by, 7, 35, 41, 42, 57, 149, 158, 159, 183n203; post–*Apollo 11* depression of, 108; religious beliefs of, 41–42; on symbolism of *Apollo 11,* 11, 42
American Civil Liberties Union, 153
American Jewish Congress, 153
American Legion, 153, 155, 210n76
Americans United for Separation of Church and State, 153
Anders, William, 39, 125; and *Apollo 8* Genesis reading, 143–46, 155–56; on Earth as Christmas tree ornament, 112; religious faith of, 40, 103, 109, 113, 143–46; on view of moon, 50, 105
angels, 1, 18, 52, 87, 89; of death, 86; in space, 48, 53, 54
Anglicans, 67–68
anthropic principle, 68
Apollo 1, 30
Apollo 4, 126
Apollo 7, 99, 133
Apollo 8, 34, 103, 113, 114, 133–34; as collective experience, 128–29, 169; descriptions of moon during, 50, 127, 133–34, 203n178; A. M. Lindbergh on, 51, 122; and reading of Genesis, 7, 8, 29, 128–29, 143–63, 167, 169; views of Earth during, 60, 102, 125–26, 133–34
Apollo 9, 98, 101–2, 108, 113–16, 149
Apollo 10, 149
Apollo 11, 11–12, 35, 41–42, 112–13, 134, 135, 149, 158; as collective experience, 89, 122, 129–30, 169; launch of, 122–25; Mailer on, 58–59, 122–23; Nixon on, 57, 58, 89; perception of moon during, 105, 111; Sevareid on, 89; views of Earth during, 102, 113
Apollo 12, 108–9, 112, 130
Apollo 13, 60, 168, 203n178; prayers for, 52, 61–62, 131
Apollo 14, 98, 104, 110, 116, 159
Apollo 15, 98, 103, 105–8, 159
Apollo 16, 99, 103, 105, 109
Apollo 17, 4, 103, 105, 125–26; launch of, 123–24
Apollo program: ending of, 6–7, 10, 74, 111, 114, 123–24, 128, 134–35, 137–38, 159–60; local social impacts of, 31–37; relation of, to environmental crisis, 3, 49–50, 59–60; religious experience and, 97–136; spacecraft, 1, 2, 5, 16, 30, 40, 97–99, 105, 143. *See also individual Apollo missions*
Apollo-Soyuz, 150, 160, 167, 170
Aristotle, 64
armed services, US, 40, 83, 88, 92; Air Force, 21, 35, 42, 85, 87; Army Air Forces, 21, 133; Army missile program, 27, 33; chaplains, 24, 147, 212n105; "character guidance" programs of, 24, 212n105
Armstrong, Neil, 34, 40–41, 58, 112, 122, 129–30, 169; on view of moon, 105
Asimov, Isaac, 124
Association of Space Explorers, 115
astrofuturism, 91, 104, 116, 124, 126, 132, 134–35
astronauts: corporate style of, 39, 95–96, 97, 99, 113–14, 169; Houston churches and, 34–35; memoirs of, 41, 99–100, 111, 112–13,

astronauts (*cont.*)
118; motivations of, 37–42; and photographs of Earth, 50, 60, 125–27; post-NASA careers of, 99–100, 104, 107, 108–9, 111, 113–21; religious experiences of, 9–10, 71–75, 99–121, 168–69; religious expression of, 26, 28–29, 73–75, 99–100, 110–21, 137–63, 167–69; in science fiction, 61, 90–91, 92; seclusion of, 5, 31, 93–95, 112–13; spiritual status of, 9, 89–90, 91–92; and views of Earth from moon, 6, 38, 45, 74, 97–98, 102, 104, 112, 113, 126, 132, 134, 144–45; and views of Earth from orbit, 96, 97, 100–102; and views of moon, 50, 91, 97, 102, 104–7, 133, 203n178; and views of universe, 73, 97–98, 102–4; workloads of, 93–94, 98, 101, 108–9, 145. *See also individual Apollo astronauts; individual Apollo missions*
astronomy, 63–66, 90, 109
atheism, 41, 109, 113, 123, 147–63; Soviet, 24–25, 89, 142
Atlantic Ocean, 20, 71, 85, 93, 100, 112–13, 127
Atlas rocket, 124
Aurora 7, 39, 96, 100
aviation, 92, 93, 127; inter-war cult of, 20, 84, 86–87, 132; and religious experience, 82–88; religious motivations behind, 18–21, 42, 43; during WWI, 86; during WWII, 20–21, 87, 165

Ballard, J. G., 92, 94
balloons, flights of, 83, 93
Bean, Alan, 5, 39, 108–9, 130
Becker, Frank, 139, 140, 147, 153
Bellow, Saul, 16, 132
Benedictine rule, 16, 17, 18, 55
Benjamin, Marina, 4
Berger, Peter, 52
Bernstein, Leonard, 128
Bible, 28, 29, 41, 60, 77, 109, 150, 155, 156; Genesis, 7, 8, 29, 65, 68–69, 128, 143–63, 167, 169; O'Hair on, 151; on plurality of worlds, 64–67; Psalm 8, 41, 57, 149; Psalm 121, 107, 162; reading of, in school, 24–25, 139, 146–47, 151, 152, 160; in space, 143, 149
Bible and Flying Saucers, The (Downing), 53–54, 63

big bang, 104, 135
Blériot, Louis, 86
Borman, Frank, 40, 85, 100, 114, 158; and *Apollo 8* Genesis reading, 128, 143–49, 155–56, 158, 207n24; on view of Earth from space, 102, 125
Bourgin, Simon, 144–46
Brand, Stewart, 125, 185n34
Braun, Wernher von, 23–24, 27, 33, 57–58, 178n107
"break-off phenomenon," 83, 93, 103
Brescia, airshow at, 86, 88
Brevard County, 43; local impact of space program on, 31–32, 35–37
Bucke, Richard, 77

Campbell, Joseph, 135
Camping, Harold, 152, 155, 157
Campus Crusade for Christ, 69
Cape Canaveral, 1, 2, 5, 31, 58, 72, 133, 164; and local impact of space program, 31–32, 35–37; views of launch from, 121–25
Carpenter, Scott, 39, 95, 96, 100
Carrying the Fire (Collins), 112–13
Carson, Rachel, 49
Carter, Jimmy, 140
Catholicism, 1, 8, 22, 57, 65, 66, 149, 150–51, 155; astronauts and, 40, 103, 109, 146
CBS Evening News, 60, 89, 123, 129, 147
Cernan, Gene, 103
Chaffee, Roger, 96
Challenger disaster, 164–66
Christian Century (periodical), 48, 69, 128, 146
Christian Crusade, 162
Christianity Today (periodical), 29, 48, 52
Christian tradition: cosmography in, 46–49, 52–54, 90, 168; flight in, 12, 18; plurality of worlds and, 64–65
Church of All Worlds, 49, 50, 51
church-state relations, 25, 29, 55, 137–63
Clark, William, 80
Clarke, Arthur C., 66
Cocoa, Florida, 35–36
Cocoa Beach, Florida, 35–37, 39
Cold War, 6, 33, 53, 88, 115; religion and, 24–26, 42, 48, 142

Collins, Michael, 31, 40, 98, 112; *Carrying the Fire,* 112–13; on philosopher, priest, and poet crew, 39, 166; on view of Earth from space, 102, 104, 126; on view of moon, 105
Collins, Patricia, 165
Columbia disaster, 166
Columbus, Christopher, 50, 80, 91
Congress, US, 24, 28–29, 61, 131, 152; Borman address to, 146; Glenn address to, 72; and religion-in-space petitions, 139, 148, 152, 153. *See also* Senate, US
Conrad, Pete, 112, 115, 130
Constitution, US, 29; First Amendment, establishment clause, 146–50, 153, 156, 157, 158; First Amendment, free exercise clause, 153, 155–56, 158
conversion, 75–79, 108, 119–20; of Braun, 23, 57; of Duke, 109; of Fallaci to space age, 2–3; of Irwin, 107, 110, 118; of Medaris, 27; of Saint-Exupéry to community of man, 87
Cooper, Gordon, 28–29
Copernican revolution, 11, 46, 63, 64, 113
cosmography, 45, 46–54, 168
cosmology: animistic, 49–51; big bang, 104, 135; and extraterrestrial life, 63–69; medieval, 25, 46, 53–54, 90, 168; in modern theology, 45, 46–49; popular, 45, 47–49, 52–54, 143, 168
counterculture, 7, 8, 69, 116–17, 125; religious, 74, 79, 109, 118
Court of Appeals, US, 150
creationism, 68–69, 120–21, 140, 155
Cronkite, Walter, 129
Cunningham, Walt, 39, 41, 99

Daniken, Erich von, 62–63
Dante Alighieri, 54
Darwin, Charles, 11, 22, 81
deism, 22, 40–41
Derham, William, 64, 66
Detroit Free Press, 153, 154
Detroit News, 154
Dickey, James, 133–34, 135
Dirksen, Everett, 139
District Court, US, Western District of Texas, 149, 150

divinity: of all, 49, 79, 81–82, 121; as ascribed to extraterrestrials, 63, 64, 66–68; technological means of recovering, 8, 16, 55; of water, 87
Donnelly, John, 159–60
Douglas, William, 147
Downing, Barry, 53–54, 63
Dreikhausen, Margret, 83, 84
Dryden, Hugh, 27
DuBridge, Lee A., 137
Duke, Charlie, 103, 105, 109, 112, 121
Duke, Dotty, 109, 121
Durkheim, Emile, 14–15, 127–28

Earth: Apollo photographs of, 6, 50, 60–61, 125–27; astronauts' views of, from moon, 6, 38, 45, 74, 97–98, 102, 104, 112, 113, 126, 132, 134, 144–45; astronauts' views of, from orbit, 96, 97, 100–102; astronomical displacements of, 63; sacralization of, 49–51, 70, 81–82, 87, 102; as site of special creation, 9, 45, 64–70, 102; as "spaceship," 59–60; threats to habitability of, 59–61
Earth Day, 50, 60, 126
education, 3, 25, 33, 53, 164. *See also* school prayer
Edwards, Jonathan, 77
Edwards Air Force Base, Edwards, California, 21
Eilmer of Malmesbury, 18
Eisenhower, Dwight, 25–26
Eliade, Mircea, 11–12, 18, 78
Ellul, Jacques, 55
Engel v. Vitale, 147, 152
engineering, 41, 43, 55–58, 84–85; NASA culture of, 30, 111–12; shallow philosophical basis of, 17, 38
environmentalism, 3, 49–51, 61, 115, 126–27; NASA and, 5, 59–60, 62, 185n34, 188n96
Episcopalians, 28, 33, 36. *See also* Anglicans
Equal Rights Amendment, 140
eschatology, 21, 45, 53, 60–61, 150, 156, 161
evangelicalism, 19, 24, 36, 73; and cosmology, 48, 52–54, 68–69, 121; high technology and, 16–17, 55–58; Irwin and, 118–21; and religion-in-space petitions, 139, 141, 155; and religious broadcasting, 141, 161–62; and

evangelicalism (*cont.*)
  religious experience, 75–79; 1970s revival of, 8, 13, 168; and space program, 25, 29, 142–43, 162–63
exploration, 13, 167; of inner space, 7, 135; of outer space, 26, 38, 50–51, 65, 89–92, 127, 129, 135, 137; polar, 91, 93; and religious experience, 79–80, 81–82, 91
extraterrestrial life, 8, 50, 52–54, 62–63, 90–91; Christian doctrine on, 45, 63–69. *See also* unidentified flying objects

Faith 7, 28–29
Fallaci, Oriana: on astronauts, 38, 96; on Braun, 58; on meaning of space age, 1–3, 4, 7, 8; on NASA employees, 30, 34
Falwell, Jerry, 140, 142
Family Radio Network, 151–52, 154–55. *See also* Project Astronaut
fascism: aviation and, 20, 87; rocketry and, 23–24, 57
Federal Communications Commission (FCC), 141, 160–62, 170
Ferkiss, Victor C., 37–38
Fletcher, James C., 152, 159, 188n96
Freeman, Theodore, 1, 2, 10
Freud, Sigmund, 78, 87
Friendship 7, 26, 36, 71–74, 79, 127, 135
Fry, Loretta Lee, 138, 148, 152, 154, 156, 157
Fuller, Buckminster, 59
fundamentalism, 24, 28, 55, 120, 151, 154–55, 156
Futurism, Italian, 84
Fyodorov, Nikolai, 22

Gabrielse, Steve, 138, 148, 151–52
Gagarin, Yuri, 48
Gaia, 51
Galileo Galilei, 54
Geller, Uri, 116–17
Gemini 4, 101
Gemini 7, 100, 144
Gemini 10, 98, 112
Gemini program, 40, 59–60, 98–101, 112, 144
Genesis rock, 106–7, 120–21
George, Henry, 17

Germany, Nazi: aviation in, 20; rocket engineers from, 2, 23, 33
Gilruth, Robert, 28–29, 144
Glenn, John, 2, 36, 71–75, 95, 100, 115; and flight on shuttle *Discovery,* 166; Kennedy and, 26, 72, 79; as model for Buzz Aldrin, 41; on ordered cosmos, 73, 102–3, 135; religious faith of, 7, 39–40, 73, 135
Glennan, T. Keith, 28
God, 17, 20, 21, 22, 26, 29, 156, 161, 169; authority of, 45, 46, 51, 55–59; care of, 39–40, 45, 51, 52, 66–70, 80, 106–7; death of, 45–46, 47, 56, 58, 66, 139, 170; fear of, 24, 41; immanence of, 46–51, 66; location of, 41, 46–54, 69, 73, 137; man as creator in likeness of, 16, 55; man impersonating, 56; man's dependence on, 9, 24, 44–45, 57–58, 102, 157, 168; ordering hand of, 73, 102–3; otherness of, 47; and other worlds, 64–69; popular belief in, 13, 24–25, 32, 51, 55; presence of, in space, 98, 106–7, 110, 118–19, 136, 168; spaceflight as commission from, 27, 157, 158; spaceflight as means of returning to, 8, 18, 58; spaceflight as prayer to, 1, 10; spaceflight as sign that man had no need for, 1, 7, 8, 109, 123; touching the face of, 83–84, 87–88, 165–66, 170; transcendence of, 46–49, 51–54. *See also* divinity; religious experience
Goddard, Robert, 22, 42
Goering, Hermann, 20, 87
Gordon, Dick, 115
Graham, Billy, 24, 33, 53, 119, 137; on *Apollo 11,* 57; on *Apollo 13,* 52; on Sputnik crisis, 25
Great Awakening, 74, 77
Great Flood, 81, 120–21
Grissom, Gus, 95, 100
Grumman, LeRoy, 42

Hadden, Jeffrey, 47
Hargis, Billy James, 162
Hawthorne, Nathaniel, 82
heaven, 52, 69, 83, 168; as inspiration for early experiments in flight, 18, 21; location of, 18, 45, 46–49, 53–54; spaceflight as ascent to, 8, 25, 42

Heinecken, Martin J., 44, 45
Heinlein, Robert, 50, 91
Herrick, Myron T., 20, 86
Heyden, Francis J., 66
High Flight (Magee), 83–84, 87–88, 107, 165–66
High Flight Foundation, 107, 119–21, 159, 165, 201n140
Hitler, Adolf, 20
Honest to God (Robinson), 47–48
Hornet, USS, 158
Houston, 2, 36, 101, 106, 113, 120, 148, 154; Astrodome, 101, 107, 118; and local impact of space program, 31–32, 33–35, 36, 37. *See also* Manned Spacecraft Center
Hubbard, Barbara Marx, 124
humanism, 52, 62, 63, 78, 157
Humboldt, Alexander von, 81, 104
Hume, David, 76, 81
Huntsville, 27, 35, 36, 57; and local impact of space program, 31, 32–33. *See also* Marshall Space Flight Center
Huxley, Thomas Henry, 54

Idea of the Holy, The (Otto), 77
immortality, 84, 87; space program's promise of, 2, 21–23, 42
Incarnation, doctrine of, 67–70
Institute of Noetic Sciences (IoNS), 116–18
Internal Revenue Service (IRS), 141, 170
International Christian Leadership, 57–58, 187n69
Irwin, Jim, 74, 99, 110, 159; family difficulties of, 106, 120; heart problems of, 98, 120–21; lunar epiphany of, 7, 106–8, 110, 118, 136, 168; religious ministry of, 75, 107, 118–21, 162, 165, 168–69, 201n140

James, William, 77, 110
Jaycees, 155; Ohio, 153; Texas, 138, 152, 153, 156
Jesuits, 65, 66
Jesus Christ, 53, 57, 78, 79, 109, 110, 118–20, 150–51; Incarnation of, 67–70
Jesus Christ Superstar (musical), 69
Jesus movement, 69
Jet Propulsion Laboratory, Pasadena, California, 22, 30

Jews, 29, 66, 153
Johnson, Rodney, W., 29, 68–69
Jung, Carl, 63, 78

Kafka, Franz, 86, 88
Kant, Immanuel, 76, 81
Kennedy, John F., 1, 26, 33–34, 111, 138; and Glenn, 26, 72, 79
Kennedy Space Center, Cape Canaveral, Florida, 31–32, 35–37, 72; "Chapel of the Astronauts" at, 158
Khrushchev, Nikita, 48
Kierkegaard, Søren, 44
Klein, Herbert, 137
Korolev, Sergey, 23
Kraft, Christopher, 28, 96
Kranz, Eugene, 27, 31, 146
Kubrick, Stanley, 8
Kuhn, Thomas, 117

Laitin, Joe, 144
Lamm, Norman, 66
Langley, Samuel, 83
Lansman, Jeremy, 160–62
Lasser, David, 91
Late Great Planet Earth, The (Lindsey), 60
Leonardo da Vinci, 18–19, 87
Lewis, C. S., 48, 65; on "seeing eye," 74, 110, 135
Lewis, Meriwether, 80
Life (periodical), 51, 99, 111, 133–34, 135, 151
Lindbergh, Anne Morrow, 51, 122
Lindbergh, Charles, 42, 51, 83, 84, 112–13; New York–Paris flight of, 19, 85, 86–87, 88–89, 93, 112, 127; spiritual convictions of, 19, 23–24, 85, 87
Lindisfarne Association, 114–16, 124
Lindsey, Hal, 60, 62
literature, 122, 132–34. *See also* astronauts: memoirs of; Bellow, Saul; Fallaci, Oriana; Mailer, Norman; Updike, John
Lovell, Jim, 100; and *Apollo 8* Genesis reading, 143–46, 155–56; on view of Earth, 60, 125, 145; on view of moon, 127, 203n171
Lovelock, James, 51
Lowry, Robert L., 37
Lucas, William R., 33

Lutherans, 23, 33, 44, 66, 155
Lynn, James, 138, 148, 153–54

MacLeish, Archibald, 50, 125, 132
Magee, John Gillespie, 83–84, 87–88, 107, 165–66
Mailer, Norman, 16, 22; on *Apollo 11* launch, 122–23, 124; on astronauts, 38, 41; on Braun, 58; on first moon landing, 58, 69, 129, 132–33, 208n38; on NASA's institutional culture, 29–30, 34
Malina, Frank, 22
Malinowski, Bronislaw, 14
Manned Spacecraft Center, Houston, 27, 28, 31–32, 33–35, 129, 133, 144, 152; mission control at, 31, 98, 102, 106, 146
Mars, 51; as dead world, 50, 64; manned missions to, 123, 130
Marshall Space Flight Center, Huntsville, Alabama, 27, 31, 32–33, 34, 57, 180n146
Marx, Karl, 25
Mascall, E. L., 53, 68
Mattingly, Ken, 99, 103
McAuliffe, Christa, 164–66
McCarthy, Joseph, 25
McCurdy, Howard E., 4–5
McDivitt, James, 40, 101
McDougall, Walter, 4, 5, 21, 172n17
McIntire, Carl, 37, 205n6
Medaris, John B., 27
Melville, Herman, 103, 122
Mercury Seven, 39–40, 72, 89, 96, 111; selection of, 73, 93–94; training of, 94–95
Methodists, 8, 27, 76, 155
Milam, Lorenzo, 160–62
Mitchell, Edgar, 74, 98, 119, 121, 135; epiphany of, 104, 116, 118, 168; parapsychology and, 110, 116–18; religious background of, 41, 109–10
Mitchell, John, 149
Moby Dick (Melville), 103, 122
moon: anticipations of first landing on, 11–12, 27, 29–30, 111, 129, 135; astronaut excursions on, 98, 103, 104, 105–9, 127, 129, 130; desolation of, 50, 60, 74, 102, 105, 127, 203n178; first landing on, 7, 41–42, 105, 129, 149, 158, 167; geology of, 98, 106–7, 108–9, 111, 133, 134, 159, 169; landing on, as national goal, 1, 5–6, 26, 28, 33, 74, 90; last landing mission to, 4, 123; responses to first landing on, 34, 57, 58–59, 69, 111, 122, 129–30, 132–34; romance of, 91, 105–7, 127, 132–34, 194n86
Moral Majority, 140, 141–42
Mormonism, 65, 159, 188n96
Mr. Sammler's Planet (Bellow), 16, 132
Mueller, George, 29, 144
Mumford, Lewis, 130
Mussolini, Benito, 20
mythology, "magic flight" in, 11–12, 18, 42

National Advisory Committee on Aeronautics (NACA), 34, 88
National Aeronautics and Space Administration (NASA), 123, 130, 131; Christian employees of, 8, 14, 42–43, 146, 167; and "citizen passengers," 164–66; and Earth photography, 125–27; and environmentalism, 5, 59–60, 62, 185n34, 188n96; and grassroots religious campaigns, 10, 137–63, 169–70, 210n76; institutional culture of, 4–5, 29–37, 95–96, 97, 110, 111–12, 169; and manned missions to Mars, 123, 159; on man's urge to explore, 79, 127, 167–68; religious attitudes of senior managers at, 27–29, 144–45, 157–60, 167, 169; and research into psychology of spaceflight, 92–95; scientific programs at, 51, 98, 105, 114, 116, 159; as secularizing force, 143, 157, 163, 167–69. *See also* Apollo program; Kennedy Space Center; Manned Spacecraft Center; Marshall Space Flight Center; space program
National Aeronautics and Space Council, 113
National Association of Evangelicals, 162
National Council of Churches, 56, 152
National Religious Broadcasters, 162
Native Americans, 49, 51, 79
nature worship, 79, 81–82, 87. *See also* neopaganism
Nazi Germany, 2, 20, 23, 33
NBC Nightly News, 50, 126
Nebuchadnezzar, 57
Nelson, Bryce, 137–38

neopaganism, 49–51, 79. *See also* nature worship
New Age religion, 75, 79
New Christian Right, 140–42
news media, 34, 35, 36, 39, 74, 79, 127–32, 160, 161; and religion-in-space campaigns, 139, 148, 150, 152, 153–54, 156–57. See also *CBS Evening News*; *Life*; *NBC Nightly News*; *Newsweek*; *New York Times*; television; *Time*
Newsweek, 59, 60
Newton, Isaac, 53
New Yorker, 128–29
New York Times, 59, 72, 111, 129–30, 146
Nicene Creed, 67–68
Niebuhr, Reinhold, 130
Nietzsche, Friedrich, 63, 66, 90
Nixon, Richard, 113, 211n81; and *Apollo 8*, 128; and *Apollo 11*, 57, 58, 129, 158, 169; and *Apollo 13*, 61–62, 131; and post-Apollo space program, 137, 159
Noble, David F., 8, 12, 21
Noonan, Peggy, 164–66
nuclear weapons, 24, 55, 56

Oberth, Hermann, 22
O'Hair, Madalyn Murray, 147–63, 169
O'Hair v. Paine, 149–53, 158–59, 161
O'Neill, Gerard K., 91, 116
opinion polls: on religion, 13, 24, 52, 78, 79, 142, 185n41, 206n17; on space, 129, 130, 131, 142, 206n17
Otto, Rudolf, 77, 110–11

Paine, Thomas (philosopher), 64
Paine, Thomas O. (NASA Administrator), 29, 146; and O'Hair, 149, 152, 153, 159, 161; and religion-in-space campaigns, 137–38, 148, 152, 155, 157
Parsons, John Whiteside, 22–23
Patrick Air Force Base, Brevard County, Florida, 72
Paul VI (pope), 57
Peale, Norman Vincent, 137
physics, new, 53–54, 104, 117
Pietism, 76
Pittenger, W. Norman, 67–68
Playboy, 150–51

plurality of worlds, 63–69. *See also* extraterrestrial life
poetry, 83–84, 103, 107, 122, 132, 133–34, 135, 165–66
Poole, Robert, 4–5, 7, 126
prayer, 30, 106, 115, 126, 148, 151, 158, 169; Glenn and, 26, 73; labor as, 16, 17, 55; recorded by Borman, 143, 149; rocket launch compared to, 1; for safety of astronauts, 35, 52, 61–62, 131, 158. *See also* school prayer
Presbyterians, 11, 35–36, 37, 41, 42, 73, 155
Project Astronaut, 138, 148, 151–52, 154–55, 160
psychology, 77, 92–95; humanist, 78; parapsychology, 110, 116–18

Rabbit Redux (Updike), 124–25, 130, 132
Rand, Ayn, 123
Reagan, Ronald, 164–65
Redstone Arsenal, Huntsville, Alabama, 32, 33
religion: cultural power of, 13–16, 32; Eastern, 79, 114; as epiphenomenon, 76–78; metaphysical, 75, 77–79, 82, 104, 117; negation of, 10, 123, 150–51, 156–57, 162–63, 166–70; neopagan, 49–51, 79; New Age, 75, 79; 1970s awakening of, 74, 135; popular adherence to, 13, 24–25, 32, 52, 55, 78, 79, 142; in post-war historiography, 7, 139–40. *See also* religion-in-space campaigns; religious broadcasting; religious experience; theology
religion-in-space campaigns, 10, 137–63, 169–70, 210n76
religious broadcasting, 141, 151–52, 154–55, 160–62
religious experience: anticipations of, in space, 88–95, 168; of astronauts, 9–10, 71–75, 99–121; aviation and, 82–88; communication of, 110–21; conceptions of, 75–79, 85; and exploration, 79–80; launch as, 1–2, 121–25; limitations on, in space, 74–75, 92–96, 97–99, 107–9, 169; and nature, 79–82; as primary ground of religion, 75–79, 111; secondary, 88–92, 121–36. *See also* conversion
Riefenstahl, Leni, 20
Rittenhouse, Bill, 119, 201n140
Riverside Presbyterian Church, Cocoa Beach, 35–36, 37

Index

Robinson, John, 47–48, 49
Robison, James, 120
rocketry, 27, 33, 61; religious motivations behind, 21–24, 33, 43
Rosell, Mervin, 24

Sagan, Carl, 135
Saint-Exupéry, Antoine de, 19, 87
Saturn I rocket launch, 1–2
Saturn V rocket launch, 121–25
Scheer, Julian, 144, 158
Schirra, Wally, 39, 96
Schleiermacher, Friedrich, 76, 77
Schmitt, Jack, 126
school prayer, 24–25, 160; Supreme Court decisions on, 29, 139, 140, 142, 146–47, 151–53, 156–57
Schweickart, Russell, 38, 41, 98, 109; epiphany of, 101–2, 108, 113–16, 168
science: and consciousness, 116–18; as justification for space exploration, 26, 29, 98, 138, 159; lunar, 98, 105–7, 108–9, 111, 159, 169; and religion, 27, 46, 51–56, 104, 120. *See also* astronomy; space science
science fiction, 50, 61, 91–92
Scott, David, 40, 98, 101, 105–8, 115, 159
Scott, Robert, 21
secularization thesis, 7, 8, 13–16, 17, 42–43, 78, 139–40
Senate, US, Committee on Aeronautical and Space Sciences, 72–73, 75, 135
Sevareid, Eric, 60, 89, 123, 124
shamans, 38, 89, 194n78
Shepard, Alan, 2, 95, 100, 104
Simons, David A., 83
Skylab, 100, 108, 114
Slayton, Deke, 2, 28–29, 39, 158
Smith, Andrew, 118, 129
Southern Baptist Convention, 33, 41, 107, 109, 110, 118, 155, 162; at First Baptist Church, Cocoa, 35–36; Irwin's work for, 119–20, 201n140; at Nassau Bay Baptist Church, 107, 119
Southern Christian Leadership Conference, 123
Soviet Union, 22, 24, 55, 142; anti-religious campaign of, 48; space program of, 3, 23, 24–25, 48, 89, 115, 167, 207n22. *See also* Sputnik
space age: dating of, 3–4, 171n15; defining role of religion in, 9, 166–70; evangelicals and, 25, 142–43, 162–63; in historiography of the sixties, 3–5; theology in, 44–70, 168
spaceflight: as a charismatic enterprise, 9; hazards of, 30, 62, 72, 92–96, 110, 130, 131, 164–66, 169; implications of, for religious thought and belief, 9, 44–70, 103, 143, 168; as religious experience, 9–10, 70, 71–75, 82, 88–136, 165, 168; religious symbolism of, 8, 11–12, 41, 42, 89–90, 167
space program, US: in histories of postwar religion, 7–9; international attitudes toward, 26, 29; public attitudes toward, 6, 9, 16, 38, 89–90, 111, 113, 129–31, 137–63; religious motivations behind, 8, 9, 11–43, 57–58, 166–68; ritualistic elements of, 15, 31, 128, 149, 157–59, 166; as secular enterprise, 10, 11–43, 123, 138, 151, 157, 163, 167–69; terrestrial applications of, 59, 111, 159; use of liturgical language in, 15, 31, 149, 158–59; *See also* Apollo program; astronauts; Gemini program; National Aeronautics and Space Administration; *individual missions*
space race, 25–26, 48, 89, 207n22. *See also* Soviet Union; Sputnik
space science, 92–95, 114, 168
space shuttle, 61, 68, 164–66, 170
Space Task Group, 137–38
space walks, 97–98, 101, 103–8, 114
Spirit of St. Louis, The, (Lindbergh), 112
Sputnik, 3, 25–26, 56
Stanford Research Institute, 116–17
Statendam, SS, 124
Stennis, John, 72–73
Stranger in a Strange Land (Heinlein), 50
Styron, William, 128
sublime, 102, 125–27; natural, 80–81; negative, 104; technological, 82–83, 123
suburbs, religion in, 24, 32, 34–35, 42, 167
sun, 63, 130; death of, 2, 61; setting of, 71, 96, 100
Supreme Court, US, 29, 139, 142, 146–47, 150–52

technocracy, 4, 12, 15, 42, 43, 59, 168, 170
technological man, 16, 30, 35, 37–38
technology, 9, 38, 42, 61, 124; failures of, 58–59, 114; implications of advances in, 45, 54–58; and nature, 82; religious concerns about, 17–18, 27, 55–57, 162–63, 170; religious inspirations behind, 8, 15, 16–18, 43, 55; and the sublime, 82–83, 123
television, 89; astronaut broadcasts from space on, 50, 57, 92, 128–30, 143–46; *Challenger* disaster on, 164–65; network news broadcasts on, 50, 60, 89, 122, 124, 127–32, 139; religious programming on, 160–62
theology: "death of God," 45–46, 47, 56, 139; existential, 47–48, 53; and lay believers, 45, 47–49, 52; natural, 76, 81; neo-orthodox, 47; and plurality of worlds, 63–69; process, 47, 49; in space age, 9, 42, 44–70, 168; of stewardship, 62, 188n96
Thompson, William Irwin, 114, 124
Thoreau, Henry David, 81–82, 84, 87, 98, 104
Tillich, Paul, 47, 49, 57, 67, 78, 90, 184n25
Time (periodical), 69, 74–75, 90, 114, 131
Titov, Gherman, 48
Tower of Babel, 57
Triumph of the Will (film), 20
Tsiolkovsky, Konstantin, 22
2001: A Space Odyssey (film), 8, 188n97

Underwood, Joseph B., 119–20, 201n140
Underwood, Richard, 126–26
unidentified flying objects (UFOs), 52–53, 54, 61, 62–63, 65, 67, 90–91, 185n46
universe, 135; astronaut perceptions of, 73–74, 97, 100, 101–4, 108, 116–17, 169; ordered nature of, 74, 90; size of, 44, 46, 63, 66, 80–81, 90
Updike, John, 124–25, 132

Varieties of Religious Experience, The (James), 77
Venus, as dead world, 50, 64
Verne, Jules, 22
Vietnam War, 3, 120, 131, 144
Voas, Robert, 95

Walden Pond, 82
Wallace, George, 140
Webb, James, 28
Weber, Max, 15, 16
Webster Presbyterian Church, Webster, Texas, 11, 35, 41, 42
weightlessness, 18, 38, 94–95, 96, 114, 116
Weiss, Pierre, 86
Wendt, Guenter, 36, 41
Westminster College, Fulton, Missouri, 50
Whewell, William, 64, 66
White, Edward, 40, 101
White, Lynn, 16, 62
Whitman, Walt, 77, 82
Wilberforce, Samuel, 54
Wiley, Alexander, 73
Wolfe, Tom, 21, 89
Woodruff, Dean, 11, 35, 42
Worden, Al, 103
World War I, 19, 20, 86
World War II, 20–21, 32–33, 87, 133, 165
Wright brothers, 19, 20, 42

Zell, Tim, 49
Zen Buddhism, 17